Is it
STRONG
Enough?

SIMPLE STRENGTH ASSESSMENTS

Dieter Wieneke, BSc, CEng, MIMechE

(retired Chief Stressman of Marconi Underwater Systems)

ISBN: 978-1-4834-4777-3 (sc)
ISBN: 978-1-4834-4780-3 (hc)
ISBN: 978-1-4834-4779-7 (e)

Library of Congress Control Number: 2016903953

Lulu Publishing Services rev. date: 05/23/2016

CONTENTS

Step Three

Step Four

INTRODUCTION

Stress and structural analysis: the art and
science of strength estimation
(or how to make sure things do not break)

It is **not** necessary for any structure to break. *Structures* means anything manufactured or built by anyone to some design or specification. It could be large or small, made of any material, or a single piece or a large collection of individual pieces joined together by any method. It could operate or be used in the widest possible selection of environments, hot or cold, wet or dry, stationary or moving.

Modern structural analysis – that is, finding the various strengths of the structure by calculation – is good enough to avoid all failures. Obviously, the structures must not be misused, but if they are properly made, maintained, and used, they should never break within their recognised and acknowledged life, if such a life has been established. The analysis may need the use of computers and elaborate programs run by experts, but nothing should ever break.

In the distant past, this was not so. It can be imagined how the ancients, in whichever civilization in the world, cursed when a large stone lintel over a doorway broke. *It was all right last time,* they may have thought, before realizing that the stone was a bit thinner this time or there had not been another storey over the earlier lintel. When they used a thicker stone, all was well, and they would have noted the necessary dimensions for success. Nowadays the necessary size can be calculated to a high degree of accuracy. Repeated trial and error to establish sufficient strength for a structure is unnecessary.

But still things break prematurely. This is irritating to a *stressman*, or structural analyst, or, the latest term, a structural integrity engineer like me. Stressmen are engineers whose job it is to make sure that **things** do not break. They can achieve this by a combination of calculation and test, a process to be described in this book. Such failures are irritating because they are unnecessary. Sorry to repeat myself, but in our times, nothing should break. Tests may have to be done as a final check on the design, strength calculations, and manufacture of the artefact, but only in important circumstances such as when safety or lots of money is at stake.

It is quite clear that often makers do not consider the strength of their product sufficiently. If they did a detailed assessment, including the effects of time and the environment, there would be much less to grumble about. Usually, however, they only call on the services of stress engineers when things have gone wrong. They have taken a chance that all will be well, that nothing breaks. If they are lucky, and they often are, especially if they have done nothing out of the ordinary, then they save time, money, and trouble; profit improves. If they are not lucky, it usually costs more to put right the failure than it does to do the job properly in the first place. It has been estimated that structural failures cost several per cent of the GDP per annum, some tens of billions of pounds sterling. This is a huge cost, especially since it is largely unnecessary.

It must be admitted, of course, that some things are well known to be fragile, and it has always been recognised that care must be taken with them. China cups are an example; there is no point in doing any analyses or tests on them. But some things are expected to be strong enough for their purpose, and it should be remembered that nowadays everyone has a duty of care for almost everyone else. In this litigious age, someone may insist on having this duty fulfilled. There are laws and regulations about many types of structures which have to be satisfied, and even where there are not, it may one day be necessary to show that attention has been paid to the strength of the item.

For regulatory purposes – say, satisfying the Building Regulations – it may be necessary to pay an expert in the field to produce a report, as the authorities concerned would not know whether to trust a nonprofessional's figures. However, if attention has been paid to the strength of the structure, this should not reveal any faults.

Costs are not just measured in money and time, interdependent quantities. Safety of life and limb may be involved; in that case, analysis and/or test is paramount. Whatever needs to be done will have to be done, and it must be done early. In established industries, the relevant authorities have usually imposed regulations to ensure safety and often environmental conditions nowadays.

There is a further cost which people do not seem always to realise, and that is to their reputations. When they appear to have failed in some respect, their critics, who will not necessarily be detractors, will think less of them. Such loss of faith may even be fatal to the careers of individuals and the business of companies. By contrast, those who correct the fault may make their reputations.

By the way, failure means that the structure no longer does its work sufficiently well. It is not necessary for there to be a fracture or an obviously bent piece of material. It may be that the structure remains apparently unaffected but merely that the safety margin has been encroached. More about safety margins later.

Purpose of this book

Many people have responsibility for the integrity of an artefact but are *not* structural experts, nor do they want to be. The purpose of this book is to help them understand their structures and then carry out simple calculations to discover their strengths and weaknesses. I am thinking of entrepreneurs, designers with many duties other than just the strength of their structures, inventors, students and their projects, and even managers who are trying to monitor the progress of people working for them. We will consider a structure of interest to you. Either it exists and you must find out what its strength is or it has yet to be designed and made.

Like managers and designers, stressmen need to know all about a structure from its birth to its death, every aspect. There may be a critical situation at any time that has to be dealt with. The strength analysis process can therefore be used as a checklist of important points in the scheme. If something is important structurally, it may be important in other respects. This could be useful to a manager.

This is all about physical structures made of wood, metal, ceramics, concrete, or plastic (reinforced or not). They are never made for their own sake. They always support or contain something to serve some other primary purpose. A teapot is a structure intended to contain tea; if it survives to become an antique and it is revered, it may be considered a work of art. So then the structure exists so that people may remember its creator or admire his workmanship. The pot now has a different role; it supports memory or admiration but not tea any longer. That would be too risky. Such a secondary role for structures, however, does not mean it's unimportant, and one of the most important characteristics of any structure is its strength.

All physical structures of any scale, from bridges and skyscrapers to watch springs and cotton threads, aeroplanes to roller blades, elephants to fleas, whether synthetic or natural, can have their strengths calculated. An approximate first estimate can be made with a calculator and only a little mathematical knowledge, merely arithmetical. If you can do numbers sufficiently well to control your budget, you can do this. You just have to get used to it. Persevere; do not let it beat you. You will be rewarded with useful knowledge. Designers will not design things that break; users will be able to estimate how far they can stretch the use of existing artefacts. If, however, even simple mathematics daunts you, then the first two chapters will enable you to describe your project to a stressman.

Readership

As mentioned before, this book is aimed at those people whose principal purpose is *not* structural analysis or stressing but whose work involves designing structures which must not fail, whether for safety or economic reasons. By following a simple procedure, most people should be able to check a preliminary design so that it is safe and fulfils its purpose. This should apply to almost all types of goods, to use the economics term – any material, any size, any shape, any purpose, any structure or part or component of an artefact or product. Even designers in a highly technical and regulated sphere producing bridges, cranes, nuclear reactors, and aeroplanes may benefit from a quick initial strength estimate.

However, designers and others must have a sufficient understanding of how structures behave and work in order to be able to make quick but not misleading estimates of their strengths, so one other aim is to give the reader some help in gaining this feel for structures. This could be described as the art bit. When you have had some practice, you will get a sense of how structures work. In particular, once you have investigated your structure, you will understand it and its limitations and possibilities. This sense will have to have two aspects. The first concerns static (and so-called quasi- static) analysis, where nothing moves, the loads being successfully supported by the structure. The second involves a trickier consideration, that of dynamic or vibrations analysis. This is often not investigated until after the first items have been produced or even sold or unwanted wobbles ruin performance or induce fatigue failure. An expert may have to be consulted.

The idea, therefore, is to give anybody the ability to estimate the strength of any structure in a safe, conservative, and simple way.

Mathematics

It is intended to describe a process – a series of steps always undertaken – to prove a structure's strength in a given set of circumstances. The process is not particularly onerous mathematically. Only simple equations will be used, no simultaneous or differential equations – only easy algebra. All the difficult stuff can be left to the mathematicians. It will be mostly a matter of substituting values into formulae.

The mathematical stuff is well documented and widely available. Every professor in every structures department in every university in the world has written papers and books giving the basic physics and maths of the subject. Their collected wisdom, in the form of equations giving stresses in particular situations, has been turned into compendia. *Roark's Formulas for Stress and Strain* (now by Young et al.) is only the best known. For a deeper understanding, I recommend the books by Stephen Timoshenko. Robert D. Blevins wrote a vibration compendium called *Formulas for Natural Frequency and Mode Shapes.* Computer programs that do the arithmetic in these books for you are available, which makes the process easy. The use of these in an intelligent way will enable anyone to ensure the strength of his or her structure, albeit leaving it somewhat heavy.

However, all these formulae which have been discovered over the past couple of centuries are based on simplified structures: beams of constant section, plates of unvarying thickness and simple shapes (square, rectangular, circular), a restricted set of edge conditions (which describe how a structure is supported), idealised or simplified loading. The result of applying these conditions to real structures which rarely display them is that the calculated stresses are approximate. This requires that the calculations must be conservative – that is, for safety's sake, they must overestimate the stress rather than underestimate it. This necessarily means the structure will be heavier than it could be if more accurate calculations were made. When more accurate results are required or the structure is too difficult to idealise by the use of simple formulae, another technique called Finite Element analysis has to be used. This needs computers, complex programs, and expert practitioners.

A word about the concept of stress; we do not refer here to the woolly, psychological state suffered by overworked people. Mechanical stress is precisely defined and takes into account two parameters: the load, including its method of application, and the shape, especially the cross-sectional shape in relation to the other dimensions, of the structure carrying the load. It is the result of small distortions in the structure which necessarily occur when a structure is loaded. There are standard equations which give the stress at any point in a structure. All you have to do is substitute the numbers for the letters in the equation, having chosen it, and do the arithmetic.

Resort to specialists or experts

This is intended to indicate when experts ought to be consulted. Obviously, the time and circumstances will vary. It will depend on the knowledge and expertise of the designer and on the consequences of breakages.

Refinements to a structure which remove significant quantities of material, or introduce a substantial change, on a subsequent design pass may need the attention of a specialist. The reasons for the specialist's involvement may be that the weight of the structure needs to be reduced or even minimised or that fatigue or creep (both time-dependent quantities) is involved, or an accurate figure for deflection is needed or a cheap material of lesser quality is being used, or just that the change needs formal stressing for regulatory purposes.

If, however, the component is not high-tech, or not completely new, a simple analysis, such as will be developed here, should be enough to guarantee its integrity. The phrase "completely new" implies structural novelty – any material or manufacturing method or shape that is unconventional in the relevant situation and therefore not proved.

Choose your expert carefully; he will only be a true expert if he has done similar work before. You need to question him as closely as you can to find out exactly what he has done. Also obtain the results of his work. If he is any good, he will be honest with you and, if necessary, suggest he works in stages or gets help. Many consultancies advertise in the specialist press. Also try to get practical, measured cross-checks. If possible, real results should always support theory.

Value

As has been said, the point of assessing structural strength is to avoid failure, which has costs in terms of safety, money, time, and reputation. Despite the existence of modern analytical and experimental techniques, there are still many design failures. The cause is likely to be insufficient attention to strength early on in the design process. It is cheaper to do some work on strength at this time. There are, of course, many other causes of structural failure.

Industrial application

Each established industry has its own methods of working; in fact, each company may have its own processes to ensure safe and sound structures. They will also have official regulations to meet if there are any public safety concerns. This applies to the petrochemical industry and the civil, nuclear, and aeronautical industries as well as to anyone who makes bridges, cranes, ships, cars, and so forth. Each uses particular materials such as steel or aluminium and whatever experience has shown to be suitable from time to time. Each develops manufacturing methods and techniques. Each uses particular types of structure, be it beams, tubes, or plates. It is a major matter for them to change any of these things.

But there are many products that are not regulated and in which some self-imposed good practices would be beneficial. It cannot be emphasised too much that early on in each project is the time to think about the required strength of the components, not when it is realised that failures will occur. It will cost much more in time, trouble, and money to put right than to incorporate sufficient strength in the first place.

Four steps

In principle, there are four tasks to consider in the strength estimation process, and it is necessary to pursue each fully to its completion; otherwise, critical situations can be missed. They are as follows:

1. The discovery of load cases
2. The setting of safety factors
3. Stress and deflection predictions
4. Comparison of stresses and deflections with allowable values

Load cases describe the life of the artefact, from deciding on the raw material with which to manufacture it to scrapping it. Anything that happens throughout its life will impose some sort of stress, and although in many circumstances this will be negligible, it cannot be assumed so. Some thought must be given to every phase of the artefact's entire existence.

Safety factors, also known as design factors, acknowledge the fact that none of the information or the calculation method is entirely accurate. Clearly this affects the quality of the calculation, and safety factors provide a margin between real stress and a calculated theoretical value which must be higher. Authorities such as the Civil Aviation Board or some parliamentary quango or working committees usually set the actual size of the factors. Their derivation can be quite complex so-called partial factors based on statistics. This sort of thing ought to be left to experts. The present work will give some simple suggestions.

Most people will consider stress and deflection predictions the hard bit. At the level proposed here, it ought to be manageable by anyone with some experience at handling numbers. The great thing is never to believe the first result; always assume that a mistake has been made and cross-check somehow.

Allowable values of stress and possibly deflections have to be taken from authoritative sources. It is lengthy and expensive and the job of experts to establish the strength of particular materials. Materials must also have a good provenance.

These tasks will be considered in more detail in the following chapters.

Summary

My intent is to help non-specialists design or use their structures so that they do not break. This can be achieved by using knowledge about their shape and situation and substituting numbers into easy equations. This process is known as doing simple hand calculations. I believe people need not be at all fearful of these equations, as they are only arithmetical.

While emphasising the value of early consideration of structural strength, I also point out that each type of artefact in each industry will settle down to a near optimum design and manufacturing technique. Change from this may be risky and expensive.

I have classified the task into four parts: discovering the load cases, setting the safety factors, performing the arithmetic, and comparing the stresses with the permissible values.

Throughout this book, I have mentioned the effect of official regulations and have tried to indicate when experts ought to be consulted.

Overall, the general idea behind the book is to determine whether any part of the structure under consideration has high stresses relative to allowable ones. If it does, experts ought to be consulted. If not, there is no need to worry.

STEP ONE

Load cases

First an explanation of the term "load case": a load, of course, is something physically forceful – a weight, pressure, an electrical or magnetic field, temperature, a twist or *moment,* a poke with a stick – anything that produces force. A load case describes a period in time, long or short, in which a load or set of loads is applied to a structure or part of it. A load case should describe a particular limited set of circumstances, a single event. It is better to have two or more cases which can be combined than to have a complex case which straddles two situations. This only causes muddle.

Section (a) Listing load cases

As indicated in the Introduction the first job is to compile a list of load cases. This should be comprehensive, from birth to death of the artefact. There are at least four reasons for making a list. First, it helps ensure that none are overlooked. If you think through the life and use of the product, stage by stage, most situations should occur to you. Second, it will be useful for future reference. When you get those horrible little moments trying to remember whether or not you have done some important job, you will be able to check back. Third, each situation can be simply defined, unambiguously and in detail. Fourth, the list may be necessary for statutory reasons.

A *critical* load case is one which produces high stresses in some part of the artifact. High stresses are what tend to break a structure. A load case does not, however, have to be critical in order to be considered.

Indeed, it may be that it turns out not to control the design of anything, but initially that may not be clear, and the fact that it has been defined does show that thought has been given to it. Therefore, it should be listed. Anyway, it may combine with at least one other load case, and that may make it part of a critical set of circumstances.

Load cases do combine. For instance, for buildings there is snow loading and there is wind loading. Clearly you can have either singly or both together, that's three altogether. Any one of the three cases may be critical. Part of a structure may have to have its dimensions designed by one load case while other parts are sized by another. Sometimes the stresses in a part due to two different cases are nearly equal. If it is required to reduce the stress by changing the action of one of the load cases, this must be remembered; the other load case's actions will also need to be changed or the the part will still be highly stressed.

It is to be emphasised that few structures experience only one load case. Usually one will be uppermost in the designer's mind, but he should pause and consider other circumstances. He may be unpleasantly surprised. Some structures – aircraft, for instance – have hundreds of cases; this situation is best left to the professionals.

Section (b) Sources of load cases

There are a number of methods of deriving load cases, some official and some not. The official ones are taken from government-sponsored organisations such as the Building Regulations, the Pressure Systems Regulations, the Aviation authorities' regulations, or many other similar regulations. A starting point for finding official regulations is the British Standards catalogue of all their publications – just follow the paper trail. Another is the Health and Safety Executive; also, the Engineering Institutions will help.

Some of these documents will specify the size of components required for normal applications. Thus the Building Regulations will give a choice of timber rafter sizes for a particular roof provided there is nothing unusual about it. No further work or calculations are needed, apart from discovering the weight of the tiles. In these circumstances, the designer needs only to find the authority governing his activities and act according to its requirements. In some cases, the loads to be applied will be given; for instance, for a floor in a particular purpose building, you are told what uniformly spread weight to use in the calculations and then what concentrated load, representing something especially heavy to apply.

Other types of regulations will give mathematical formulae which are to be used. In the Pressure Vessel code of practice, for instance, the required vessel wall thickness is calculated directly from an equation. Obviously, you first have to know the maximum pressure, which the user must supply, and decide on other dimensions, material, and so forth. In these cases, some of the work is done for you. With minimal effort, you are given a safe thickness for your structure. The other dimensions of the vessel will be dependent on such considerations as the space available or the component's capacity. Obviously, you could find the allowable pressure if you had an existing vessel; you do the calculation backwards by manipulating the equation – a bit of simple algebra.

Sometimes the regulation will be more basic and you have to do some work to arrive at a load. The aviation industry has a "flight envelope" which is presented as a graph of aircraft speed against acceleration. All situations within this envelope must be catered for. Obviously, the corners of the envelope represent extreme conditions. The top right-hand one gives the maximum speed together with the worst acceleration – pulling out of a high-speed dive. Often this will be a highest load case and must be used in assessing the strength of the various parts of the aircraft, but not always. Sometimes an intermediate situation will give larger loads, perhaps because another load case is also active in that situation and adds to the flight load or because the load direction and therefore the direction of the acceleration may be important.

Often to these official standard derived cases you must add loads which are special to your particular structure. You might have a very heavy machine that you want to support on the new floor of your loft. This will add to the uniformly spread load specified in the regulations. You will also have the standard concentrated load case to consider, which might be another unexpected concentrated load in addition to your special one. Instead of normal rafters, you will have to design especially deep ones or put in an RSJ (rolled steel joist).

These official regulations or codes of practice are always based on existing technology and usage so that the structures they control are essentially similar to previous designs – the only differences are in simple matters like size. New structures will be within the limits set by previous designs; nothing is significantly different or original. The codes never anticipate new trends; they could not. When novel features in the technology arise, the writers of Standards have to modify it. But the inventor, not officialdom, starts the process.

If there are no standards to help, you will have to work out your own load cases. First and foremost, you should consult the specification for the product. If you are lucky, this will exist and will give detailed, numerate rules for the performance expected from the finished article. But maybe you haven't written it yet. Now is a good time for a first draft because it is time to specify numbers. Load cases need numbers which define the size of the loads. From this the sizes of the structural part of the components will become clearer. If these are unsatisfactorily large from the point of view of other design aspects the design will have to be rethought. Detailed parts of structures are often undersized; usually this can be put right without compromising the project very much, although individuals may grumble because their part is affected. If you have a contract to do this work then it will be a source of the information required, at least qualitatively. If it is your own project, you will have to invent or somehow derive the required data.

Incidentally, you must also have drawings of the structure or at least a sketch, an idea of what it looks like. It is amazing how often this is overlooked by inventors or entrepreneurs. This happens because they concentrate on one aspect only, perhaps the electronics. Eventually they have to become more specific. So you might have to do some drawing or at least sketching as well. This sketch will need some dimensions.

In order to cover all the possible cases, you will have to construct a comprehensive through-life history of the artefact, from manufacture of the structure to disposal. It is perfectly possible that the worst stresses occur during some manufacturing stage and not in normal use at all. This can occur in the construction of some bridges when a central section of the span is hauled up to complete it. Also, disposal can produce critical situations – for instance, in prestressed concrete beams which have their reinforcing bars tightened as the building develops. At completion, the bars are cemented into place permanently. When it comes to demolition, it may not be possible to loosen them progressively, and as the building comes down, the load reduces and the bars bend the beam in which they are embedded to the point where the beam fractures, perhaps explosively.

If you are working in a team designing some structure, you will have to keep in close contact with the other designers so that you understand every nuance of every change in the design and alter or add to your load cases to match. You may happen to overhear some designer or boss talking about his desires for the project, which could signal a new load case. You have to be on guard. Others can increase your problems on a whim.

You may be asked or wish to do a check on an existing structure, perhaps with a view to changing its use or increasing its capacity. You may not have any design calculations or a formal load case list. You can extract the information from any paperwork, such as operating or repair manuals. These will tell you what the thing is meant to do, from which load cases can be deduced. Repair manuals often have those exploded pictures of the components, and the list of components might specify the bolt grades which will give you their strengths. You can work out the maximum loads expected to pass by.

Section (c) Load types

Apart from loads specific to the actual purpose of the structure, maybe to hold up something in a building or hold down something on a vehicle or anything else, there are a few general matters to consider. Not all these points will be relevant in every application. Many of them are not obvious but may lead to significant results.

1. Mass

Mass is a prime provider of loads on structures. By the way, weight is a measure of mass; it is the force due to mass, usually in the earth's gravitational field. The words are sometimes used interchangeably, but they should not be. Sometimes mass is the largest contributor to the loads on a structure; sometimes it is negligible. Examples of the former situation occur in civil engineering, buildings, bridges, and so forth. The latter may occur in components of structures such as vehicles where other forces are much larger.

The effect of mass can be magnified by accelerations, both linear (or straight line) and rotational. This is the expression of Newton's Second Law, that force equals mass times acceleration. Anything that is moving must have suffered acceleration, and if it changes direction or stops then it suffers some more. These accelerations can be very large, thus generating large forces. For instance, when a car goes round a corner sideways forces are applied at the tyres. Even if the passengers do not feel it very much, the suspension links will.

While still on the subject of mass imposing loads on structures, it is pertinent to mention a quick and easy shortcut to the establishment of a set of design loads on certain types of structure. These are known as handling equipment and apply where a mass is to be supported or moved. The idea is to take the mass of the supported item and multiply by four. Two of this four is to protect against dynamic effects, the other two for unknowns. The stressing calculations are then performed at this artificially high loading to guard against failure in the ultimate case. This term will be explained later.

If the structure is involved in lifting the factor should be six. A beneficial point is that the actual stresses are usually below the fatigue threshold (see later). However, note that there may also be other important loads to consider in addition; the above method may only give some of the cases. This load is downwards due to gravity. Lesser loads acting horizontally are also specified, viz. the mass times one in the forward direction and the mass times two in the sideways direction. Sideways is taken to be in the shorter horizontal direction, forwards in the long one; obviously, they can be in either direction. This is not a logically derived set of loadings. It is justified by long experience of the conditions experienced by such structures; use it and you will usually get a sufficiently strong and durable device.

2. Rotation

The other effect of rotation is to produce a force towards the centre of rotation. For example, if a conker on a string is spun around and around, the string is pulled taught into tension so there is a force in it. It must be remembered that rotating bodies have this force within them. It is proportional to the speed of rotation *squared* – that is, one has to multiply by this velocity twice. This means that it becomes more and more important as the speed increases. The Millennium Wheel designers clearly did not have to worry about it at all. Engine designers do.

3. Pressure

Another source of loading is something that must be treated with the greatest care, and that is pressure due to fluids. Fluids are either liquids or gases. Pressure is due to each molecule or atom of the fluid bombarding the surfaces of its container so that the load is applied continuously all over the surface. Atoms and molecules are always moving except at zero degrees on the Kelvin, or absolute, scale of temperature. The trajectory of the molecules or atoms is obviously random so that they strike the surface at all angles. This means that only the ones hitting the surface at exactly right angles have any effect on the container; for any atom or molecule coming at an angle, there is always another one at such an angle as to cancel the effect of the first. Therefore, pressures are always applied at right angles, or *normal* to a surface.

There are regulations concerning the operation of pressurized equipment issued nowadays by the EU for all systems carrying more than half a bar (a bar is the pressure due to the weight of the atmosphere above). Even half an atmosphere needs thinking about since the load due to a pressure is found by multiplying the pressure by the area affected. Half a bar acting on a standard door, for instance, produces more than eight tons of force. This will certainly burst it open!

The thing about pressurized fluids is that they get everywhere unless positive and effective measures are taken to stop them. We all know this: a tiny amount of spilt milk goes everywhere, and that's only under the influence of gravity. This means that a careful survey of any design containing fluid must be made to find out where it will penetrate. If it goes round a corner it will start pushing in a different direction. The trick is to install seals, as it is usually necessary to take positive steps to stop fluids flowing through even very small gaps. It is then obvious how far the fluid will travel and therefore how much area is being pushed. Remember that there may be a pressure on the opposite surfaces of your component, either inside or out. It is the difference between the pressures across any surface that gives the net load.

For instance, think of one of those big tanks in a chemical works, a huge cylinder standing on end. If it is full of some liquid, then it supports a considerable pressure, especially near the bottom. Pressure in this situation is the height of the liquid surface above the bottom times its density times the acceleration due to gravity, if you remember your fluid mechanics (if you ever studied any). This pressure acts outwards. Now, if this tank is out in the open, it will sometimes also be in the wind, which could be a gale. The windward side of the tank wall will be pushed inwards, and this pressure counteracts the liquid's pressure. Therefore, the wall is relieved of the necessity to support all the liquid. The wind does part of the job, but only temporarily. On the leeward side of the tank, the pressure due to the wind may become a suction, and now the wall has to support more pressure than just that due to the liquid. This is more important, of course, and the wall will have to be thicker. If there is a hatch, the fixing bolts will have to be stronger.

The atmosphere, which is miles deep, exerts its own pressure, but since this usually gets all around any object, inside and out, it exerts no net force on it. Normally pressures are quoted relative to atmospheric pressure and its value can then be ignored. Only if vessels are pumped out, creating a partial vacuum, do you need to remember it. Pressure gauges measure relative to atmospheric pressure, which is useful.

Talking of the wind, there is a pressure due to motion, as when you put your arm out of the car window when it is moving – the so-called dynamic pressure. If something moves through a fluid, a pressure is exerted on it. This is due to the density (it's much harder to walk through water than air) times the speed of travel, which has to be squared (which means, remember, that you count that twice). Another factor might be called a streamlining effect – it's more difficult pushing a plate through water face on than edge on. The speed is most important, though; if it is doubled, the load is quadrupled. If the speed is trebled, the load goes up nine times – and so on.

A more obscure effect occurs when objects accelerate in a fluid. It becomes important in more dense fluids, water, or syrup. If you wish to accelerate the object, you must also accelerate the surrounding fluid. This is the "added mass effect", and in the circumstance that your product suffers this effect and it turns out to be important, it is better to consult the experts.

4. Temperature

Temperature effects should be considered. Temperature change does two things to a structure. Increasing it usually causes the strength of a material to decline; how much and at what level is entirely a property of the material. Decreasing it increases strength, although not to the same degree. Notice that the change from the manufacturing temperature is what matters. The other thing is that most materials expand with increasing temperature. If a component is restrained from this expansion in any way, it becomes stressed, and these stresses may add to those existing for other reasons.

Loads can be produced by temperature differences. This difference can be a change of temperature in the component over time, or a variation of temperature in different parts of the component, or both. Consequently, the structure changes size and shape, and it is this that produces stresses and strains. It is important to discover what the restraints are on the component, if any, as these control the stresses and their position. For instance, a bar which is heated up to the same temperature all over but is just lying on the ground has no stress in it; if it is trapped between two solid walls when cold and when heated tries to expand, it cannot as much as it wants to and thereby develops a compressive stress.

5. Deflections

Some parts of a structure are stressed because they have a deflection imposed on them. They do not have a direct load such as a mass applied to them. Of course, somewhere in the structure, there may be a load, but in the part of interest, the only sign is some distortion. It may be possible to calculate the stress in it from deflection measurements. An example is an interference fit between a pin and a hole slightly too small for it; another is tightening a nut and bolt to a degree greater than just enough to hold two things together. If the relevant sizes of the items are known, the stresses can be calculated. This effect can be treated in a similar way to thermal calculations.

6. Vibration

Then there is the problem of vibration. All structures have so-called natural frequencies; that is, if they are suddenly disturbed from their resting shape they do not just return to the original shape and stay still. Instead, they zing back and forth until they have dissipated the energy imparted by the disturbance. They do this a certain number of times a second, which is called their natural frequency. This is controlled by the shape, size, and material of the structure. However, the alternating stresses engendered by this mechanism can be damaging by causing fatigue, of which we will discuss more later. Vibrations can be controlled by dampers or by changing their natural frequency to a harmless level.

They can get out of control. This happens to a structure when an alternating force is applied, that is, one which changes in size and/or direction over time. If the frequency of the force variation nearly or exactly coincides with the natural frequency of the structure, deflections build up and the stresses can be high enough to cause failure.

So the load case is provided by a comparison between the natural frequency of the structure in question and any so-called driving frequencies. These are the frequencies due to the alternating loads. They may be regular, such as the piston in an engine going up and down, or random, such as the wheels of a car rattling over a poor surface. There are numerous natural frequencies in any practical structure. Mostly it is the slower ones that are important, but if the driving frequency coincides with a higher frequency, then that is the one which may get excited and do damage.

The calculation of stresses due to vibrations, if they cannot be avoided, is complicated and must be left to specialists. Merely alternating loads, which do not cause the structure to vibrate at one of its natural frequencies and so become uncontrollable, may cause a peak stress that is high enough to start fatigue damage. This also becomes complicated enough to require specialist attention because a life to failure will have to be calculated, although if the peak stress can be kept low enough, the life will be so long that fatigue can be ignored.

Varying loads are either of short duration or continuous for some relatively long time. The longer duration ones are either regular or random in time and/or amplitude.

6 a) Long-term regular dynamic forces

These are like the waves on an otherwise still pond just after you have dropped a pebble into the water. Their *amplitude*, or height from trough to peak, and their length, from peak to peak, are unvarying. The time for each wave to pass a given point does not vary either. These loads are the easiest to deal with mathematically because they are so regular. Their variations are usually taken to be sine waves or some similarly regular wave form. Load cases which exhibit these forces have to be considered because their *frequency* should not coincide with the *natural frequencies* of the structure to which they are applied. If they do, then the structure deflects under their influence without limit, theoretically anyway. The structure will fail. Frequency is defined as the number of times the loads occur or the structure deflects per second. I will explain the fatigue connection later.

6 b) Long-term irregular dynamic forces

These are irregular in their amplitude, length, frequency, or any combination of these. Often the *waveform*, that is the variation in amplitude with time (this is usually plotted on graph paper) is the sum of a number of regular waveforms and these can be extracted by the use of specialised techniques such as Fourier transforms, but we will not go into that here.

6 c) Short-term shock and impact loads

Think about the effects of *impacts* or *shocks*. Technically these are very large loads which act for a very short time. This is difficult to deal with analytically in a simple way, and the best method for avoiding breakages is to design the component similarly to one of proven robustness in this sense, which may include protection methods such as using boxes or padding. It may be necessary to do tests on your new product. The problem with the latter course of action is that it is expensive. Moreover, if a weakness is detected, it will be even more expensive to remedy, as the test will have to be repeated. If shock loads are likely to be a problem, it is preferable to consult specialists, who will reach for their Finite Element computer package and use its transient facilities, which may avoid tests.

7. Moving loads

Another effect that transgresses the slow load application rule is that of a moving load. If a bus crosses a bridge, the bending in the bridge is partly dependent on the speed of the bus. This all has to do with the natural frequency of the bridge in bending, and if such a critical case arises in the design of your new artefact – that is, it must resist a moving load – again, it would be advisable to consult an expert.

8. Magnetic and electrical effects

The next consideration is that magnets and electrical currents flowing in wires produce forces which may be large enough in some artefacts to cause concern.

9. Manufacturing processes

Another rather insidious cause of stress in apparently unloaded structures is due to the manufacturing process. If the component is made of metal, it may have been heat treated – heated and then quenched by plunging into a cold liquid – in order to develop the material's strength. This can cause what are known as residual stresses, due to uneven cooling of the metal. Plastics also have this problem. Here you need to know what the manufacturing method is or is going to be.

This residual stress often shows up through unexpected fractures, even though the component was otherwise properly analysed. It can be measured by an experimental process involving strain gauges whose centres are drilled by a small drill bit. The change in stress is due to relieving the residual stress. There are specialist companies who do this. Different manufacturing methods may be called for.

10. Failure cases

Failure cases must be considered. If part of a structure breaks, other parts of the structure will probably have to carry a bigger load. For safety's sake, it may be necessary to ensure that there is no further failure. Clearly higher stresses occur in the remaining structure, and they should be sized to cope. If a lower safety factor (see later section) can legitimately be used, for instance, by limiting the time the broken state is tolerated, then there may be no change to the structure.

In general, single failures are considered. Unless there is a complex requirement, which is best left to specialists, multiple failures are not normally demanded. The failure case must include consequent dependent failures. If one breakage leads to another, the second must be included in the calculation. There may be a number of failure cases because more than one part of a structure may fail independently.

11. Misuse

Lastly is the matter of reasonably foreseeable misuse: if the gismo can be misused in some way, then sooner or later it will be (First Law of Universal Cussedness). The more items are sold and the more widely they are used, the more likely misuse will occur. Indeed, the owner of one gismo may well be pleased with his new way of using it; he may find it very helpful, while you, the designer, may cringe. This may impose a different set of loads on the structure and constitute another load case(s) which must be taken into account. The word *reasonable* obviously needs defining. Dropping a television out of an upstairs window is not reasonable, and the casing need not withstand it, but putting a pile of heavy books on top may be. The casing should not split open, exposing high voltages to the touch. As a last resort, it may be necessary to warn against some misuse by means of a label or a note in the manual, but this is not the best practice and should be avoided. It is best to design to avoid misuse if possible, though this is a design issue, not strictly a strength one.

Section (d) Summary

A load produces forces in a structure, and there are many sources of loads, depending on the circumstances. It is important to list the cases in a formal manner to make sure nothing is overlooked, for future reference and to enable each case to be simply and completely defined. It may also be required for regulatory reasons.

Sources of load cases come from official bodies, or they may have to be derived by the designer himself. This requires an imaginative, thoughtful trip through the life of the component or product from its initial manufacture to its final disposal, which may not be just to hurl it into a dustbin but may itself be a critical operation involving strength considerations. It is not unknown for a component to experience its most severe stress condition just once in its life. Indeed, it may never actually get into that state. However, structures are designed to handle the worst credible thing that can happen to them. Anything less and there will be a danger of failure. However, do not bother with anything that is incredible; it is just as important not to overburden the design with impossible conditions as to make sure it can cope with possible ones.

It is prudent to examine every possible issue just in case a severe operating condition for the product lies hidden within it. The designer who is considering the strength of his artefact must know all there is to know about all the circumstances in which it could ever operate or exist. As the project progresses, new load cases may arise. Either new knowledge or new requirements are introduced. The designer must not be afraid to reconsider the structural strength of his artefact when this happens, even though it means some redesign.

Twelve load types were identified. They are listed below:

Number	Type
1	Mass
2	Rotation
3	Pressure
4	Temperature
5	Deflections
6	Vibration
7	Moving load
8	Magnetic and electrical currents
9	Manufacturing residual effects
10	Failures
11	Misuse

It must **not** be assumed that the above list is exhaustive. It is just the list of conditions with which the author is familiar. There may well be other circumstances producing loads, although these should cover most.

STEP TWO

CHAPTER 2

Safety Factors

Whatever is a safety factor? It's a number by which you increase the loads or decrease the strength of the material. This little trick gives an extra margin between the actual strength of a structure and the loads imposed on it. Why? It is because you do **not** know anything exactly. In practice, nothing can be measured or performed or created exactly. Everything is subject to inaccuracies, and although nowadays these can be reduced to astonishingly low levels, it is expensive to achieve high degrees of accuracy. The less certain the various parameters of a project are, the larger the safety factor ought to be.

There are five categories of variability leading to these uncertainties. These are: load levels, calculation accuracy, material strength, manufacturing variability, and service practice.

Section (a) Load levels

Although the worst credible loads may have been calculated with as much care as possible, there is always a doubt about them. Often loads are found by measuring them in tests in a laboratory or in field trials. An average and maximum are found and can be used as the basis of the case. Obviously, though, there is the possibility that the trials were not quite thorough or representative enough, which introduces a possible error. It could be that some basic fact is not known precisely. It has to be estimated at least to some extent, perhaps leading to more errors. Maybe a particular cargo is a bit heavier than usual, a weighing machine is not properly set, or some other load source is unexpectedly out of the ordinary.

Perhaps you did not know some significant detail. In one instance, in the author's experience, the maximum depth in the sea to which a submersible would ever go was stated unambiguously, so the sea pressure was known – no argument. Sometime later, it emerged that the sensor which controlled the vehicle's depth was only accurate to plus or minus a few per cent. The submersible could easily have gone that much deeper. If something had been only marginally strong enough, that could have been disastrous. A Safety Factor saved the situation.

Note here that safety factors do not allow for new, unaccounted for or unexpected load cases or the use of the artefact in a way not provided for by the designer. They only protect against lack of precision – and then only to an extent limited by their value. It is still essential to estimate loads as accurately as possible.

Section (b) Calculation accuracy

Perhaps your calculations are a little wrong. Mathematical equations are often derived and made solvable by approximating the situation they describe. Usually some items in the equation are ignored on the grounds that they are not very big compared to the others, or it could be that the method of solution is approximate. These manoeuvres are justified because they have given good answers in the past. As time passes and structures evolve to meet new circumstances, these approximations may no longer be quite so true. If the drift from accuracy is not too drastic, the safety factor may protect you from disaster. Of course, this should only happen when you are not aware of the questionable applicability of the equation. The safety factor safeguards you against your ignorance. If you know about the discrepancy, you should allow for it specifically; the safety factor is to cover unknowns. It should never to be used to deal with known situations.

Then there are plain arithmetical errors, such as $2 + 2 = 4.5$. Of course, the mistake may be so large that your safety factor is not adequate. This means you must do checks on all your workings in order to find such errors; it may be lucky if the safety factor protects you.

Similarly, if your input data – geometry, loads, and so forth – is wrong, then your results will be wrong too. Again, try to check in some independent way.

Section (c) Material strength

Materials are a source of uncertainty. Worked or heat-treated metals are most consistent in the strength that they are predicted to have. This is because a so-called minimum figure is used. This is calculated from a large number of specially performed tests, which guarantees that most, but not all, samples of the material have at least the specified strength. It can be better by 20 per cent or more, and this in itself is an added but generally unknown extra safety margin. This illustrates the fact that to avoid uncertainties in the structural integrity sense, some inaccuracies have to be tolerated. However, they are in the conservative direction; the structure is stronger, not weaker, than estimated.

Working metals means forging, rolling, or extruding them – in general, giving them a good bashing. This breaks up the large grains which make up the pigs, or blocks, cast from the molten state. Heat treatment gives a similar effect. In these states, the metals are usually stronger. If they are not worked or heat-treated, they are often brittle, as are many other materials. These require extra safety factors.

Section (d) Manufacturing variability

It is not possible to make anything perfectly, that is with no error in the length, breadth, or thickness. Even if by luck you did get something exact, you would not know it because you could not measure it exactly either. So you have to cope with dimensions which are a little uncertain. In practice, most things are made sufficiently well that the stresses do not vary significantly due to these small variations in dimension. There are, however, a few exceptions.

Where a load is transmitted by one narrow component bearing, i.e. leaning, on another narrow one, the minimum and maximum thicknesses may be critical. The machining tolerance on a thickness is likely to be one-fourth of a millimetre, however thick the component. If this has a thickness of similar size, the tolerance is obviously significant – you must do the calculations using the thinnest value. Again, if the failure to be guarded against is buckling of a thin sheet, it is necessary to work on the least thickness, as the strength is proportional to the thickness squared or more. The buckling failure of the component will be sensitive to the thickness.

Section (e) Service practice

The hope of every designer must be that his product is used exactly as he envisages. This is a vain hope. Therefore, he must try to ensure that either he physically prevents misuse (ensuring that components only fit together one way) or that it is not harmful in any way. In some cases, this is done by writing careful procedures. For the aircraft industry, this should be acceptable. It is a highly developed, well-ordered, and disciplined industry. Everyone is well trained. Everything should always be performed according to the instructions, shouldn't it?

This did not work out in one case in the author's experience. Jet-engined aircraft fly at high altitudes because this is where they are most efficient giving the lowest fuel consumption. The stressmen obviously assumed that the aeroplane would fly most of its life at a high altitude. However, in this instance, the pilots of a particular airline could not be bothered to climb to more than half the recommended height, as their journey was quite short. The problem with this concerned the wings. One of the load cases for the wings is due to gusts in the air through which the aeroplane flies. Obviously, the wings flex a little extra, and this diminishes the fatigue life of the structure. Gusts are fiercer at lower altitudes because the air is denser, and what the pilots were doing was shortening the aircrafts' lives. When this was discovered at the next regular inspection, some hasty recalculations had to be done!

The example shows that it is possible to misuse a product either inadvertently or deliberately but arguably justifiably (nobody said not to). This could be a load case inadequacy or the user's fault or both.

As in the case of calculation accuracy, the safety factor will only guard against small load variations. It was enough in the above example.

Section (f) Safety factor types

Because there are many failure modes, which I will cover in more detail later, there are as many safety factors. For instance, ductile metals can fail by actually fracturing so that a split, a jagged edge, appears. The component may fall into several or many pieces – this is known as an *ultimate* failure. However, long before such an occurrence, the metal may stretch irreversibly – that is, long beams will bend permanently, the body of a car will dent, and so on. This is a type of failure; it is where the metal is said to *yield*. Misshapen components usually cannot do their job properly. There is no need for fracture to occur to say that a structure has been broken. We therefore have two safety factors, one for yield and another for the ultimate case. It is usually true to say that an ultimate failure is more severe so the factor is larger than that for the yield condition. Brittle materials are different; the ultimate and yield points are very close or identical. Here it is usual to have an additional factor – for instance, a so-called *casting factor* for cast and unworked materials.

Fatigue is a more difficult type to deal with. Again, there need not be any disintegration of the structure since it is sometimes possible to see a crack early on in the life of the component and even follow its progress without the structure becoming unusable. A branch of stressing called "fracture mechanics" deals with the progress of cracks so that a component can be taken out of service at a suitable time. This is beyond the scope of this book.

The ratio between stress and the time to fracture (known as the *life*) of a fatigue crack is not linear. If the stress in a component is doubled, the life is less than half that at the lower stress. Conversely, halve the stress and you may make the life not just twice as long but infinite. For this reason, the safety factor is not applied to the load or stress but to the life. Obviously, one quotes a lower life than the calculated one. This implies that a history of usage must be kept and written down so that the life is not exceeded, perhaps dangerously. If this is expensive, unlikely to happen, or just a bore, it is better to design to what is known as the *fatigue limit*. If you design to this low level of stress, then fatigue will never occur or will only occur long after the artefact has become useless. The precise details are material dependent and will be described later.

A similar situation occurs with the phenomenon of creep. This is a time-dependent failure, so the safety factor is applied to that, not the stress. This type of behaviour and its inverse *relaxation* is later described in more detail for various materials.

Section (g) Suggested values of safety factor

These are dependent on material and intended use and quality of manufacture. Some materials are more reliable and consistent in their behaviour; some usages are potentially dangerous; some manufacturers or their methods are variable. All these matters have to be taken into consideration when deciding on the safety factors.

Table of suggested Safety Factors

Material type	Some typical materials	Application	Controlled engineering		General engineering	
			ultimate	yield	ultimate	yield
Ductile materials	Most metals and their alloys (such as steel and aluminium alloys)	Lifting gear and lifted items	4	3	6	4
		Safety related	2	1.5	3	2
		No direct safety implications	1.5	1.25	2	1.5
	Unreinforced plastics	General use	3		5	
		Safety related	6		10	
Brittle materials	Metal castings An extra factor to those above	all	1.6	1.6	1.6	1.6
	Ceramics				4	
Composites	Plastics with designed reinforcement (longitudinal and woven)	Advanced applications	1.5			
Composites	Mats with random directed fibres, short fibre doughs, and so forth	General engineering			2.5	
Any	all	Fatigue stress > fatigue limit	Safety Factor on life **5**			
		Fatigue stress < fatigue limit	Safety Factor on fatigue stress **2 to 10**			

Notes on the Table

1. The suggestions in the table are quite high. If this causes problems and a lower value is decided on, more care must be taken with the strength estimation process. Perhaps it is time to consult the experts.
2. It is, of course, possible to have different safety factors for different load cases, even if they are subsequently combined. There is a possible limitation to this when the cases or situations are non-linear.
3. In some circumstances, the SF should be higher than normal – for instance, when something heavy is lifted.
4. A further distinction which can be made is between what I call general engineering and controlled engineering. In the former, it is likely that no more than a passing interest is taken in the strength of the structure; as long as it does not break, nobody worries. It could easily be much stronger than necessary, but that does not matter. It might be that the structure was just about to break in the worst circumstance, but that too is unknown. In the latter case, it could be that weight is very important (as with aircraft), safety is vital and authority is imposing care, or both. In any case, the stresses in all cases have been calculated and critical ones tested. This is a professional procedure and outside the scope of this book. If in doubt about the status of the artefact with respect to any of the uncertain aspects listed above, assume the worst and use the general engineering figures.
5. The most obvious safety issue in ordinary life is dropping heavy items which are being lifted. The lifting gear is usually subjected to the given Safety Factors. Other safety-related items which may be less important, perhaps because they are merely contributory to the overall safety, may have lower factors. Lower factors still are acceptable for most structural components.
6. In general, unreinforced plastics should only be stressed for short time periods; they have high factors because they are of variable quality. Their quality is dependent on the manufacturer and his methods. They are a specialist subject.

7. Metal castings are a special hazard because the casting flaws are not removed in the production process, and these cause stress concentrations which reduce the strength of the artefact. An extra SF is used; the load is multiplied by the usual factor and by the casting factor. So a general engineering ultimate factor of 2 becomes 3.2.

8. Ceramics suffer from the same problem – flaws. They are basically strong, but again, the stress concentrations at a sharp flaw combined with the fact that they have very low elongation to failure make them tricky to use. A high factor is recommended; also try to design the item so that the stress is compressive, if this is still possible. This closes the flaws, which will then not grow larger.

9. Reinforced plastics, or composites, come in two types as far as RFs are concerned. The first has a carefully designed, high-volume fraction fibre lay-up produced with advanced manufacturing techniques. This process produces a reliable product, and the usual low factors can be used. These structures are specially designed, analysed for strength, and produced. The second type is not so refined and carefully designed. There is the lay-up which apes metals – that is, the fibres are randomly orientated so that the strength is the same in all directions; there are doughs with short fibres, also randomly directed, and so forth. These are more general engineering methods, not so reliable, so they need a higher factor.

10. Most common composites use plastics which are relatively brittle, so only the ultimate factor is given.

11. If the structure has a critical fatigue case and a life is calculated, then a safe thing to do is divide that life by a large SF – five is suggested here. If that is a problem, expert advice should be sought.

12. If the fatigue limit stress is used as an allowable, a lower factor can be specified. Depending on the material and circumstances (manufacturing and safety) SFs between two and ten are suggested.

General notes

An important point is that only one set of safety factors should be used. If several people are involved in finding the loads and doing the calculations, the temptation exists that they will all add their own margins just for luck. They may round up the figures at many stages. They should try to be as accurate as possible. If the processes are not disciplined, the result will be that a safety margin is imposed which is too large (expensive in various ways) and unknown in size and therefore difficult to deal with. The project will be unnecessarily burdened.

Different industries do the safety factor job in different ways. The suggested method in this book is the simplest, but it is possible to split the overall safety margin between strength and load into a number of factors. All the constituent parts of the design-to-use process could be individually assessed and then the resulting factors combined. This should certainly give a more accurate overall safety factor which may or may not be smaller, but it implies that detailed knowledge is available for all the processes. This is not for the non-specialist designer to whom this book is addressed.

Often the allowable stress is divided by the SF and the actual load is applied in the calculations. For *linear* structures (covered more later), this is mathematically identical to the method suggested in this book. Strength values given in the literature are then lower than the yield or ultimate strengths. This may leave the value of the SF unknown.

The stressing method used in this book is known as the deterministic method of stressing. This uses SFs as described above. There is a probabilistic method. This assesses the probabilities of the strength of the material, the occurrence and level of the loads, and the severity of the usage – indeed, anything else that is relevant. A view of an acceptable probability of failure has to be taken, and then the calculation of actual failure probability can be done and compared to the acceptable. This method requires much statistical data, which is often not available; it is still being evaluated (at time of writing) and certainly requires special expertise. The hope is that sufficiently strong structures that are also lighter and/or cheaper can be produced.

Section (h) Summary

Because of the innate inaccuracy of all the processes of producing anything, from first gleam in the designer's eye to final use, it is necessary to have a safety factor. This involves artificially increasing the value of the loads by a considered factor.

There are five sources of these inaccuracies: load levels, calculation accuracy, material strength, manufacturing variability, and service practice. If in your case you are more than usually uncertain of the reliability of one of these categories, you can increase the safety factor.

Some safety factors are applied to the loads on a structure, some to the calculated life. There are other more complex methods of proceeding.

A number of conservative, suggested values of Safety Factor have been given.

STEP THREE

Stress and deflection predictions

In the next few chapters, the methods and procedures required to find the stresses and deflections in any structure will be described. These stresses will be approximate but somewhat larger than in reality and should give a good idea of the necessary size and shape of the structure. Alternatively, if the structure already exists, the loading it can carry can be reasonably estimated.

So far, we have considered the load cases, which describe its life. Many of these you can neglect because they are obviously, in all circumstances, less severe than others. What we need are the severe ones. Structures must be designed by the worst credible cases, as has been said before. We have also decided what safety factors to apply to these loads. So now we have a good numerate idea of the severity of the duty the structure has to bear.

There are five steps in the stress and deflection prediction part of the stressing procedure:

1. The discovery of all the necessary structural data (Chapter 3)
2. The classification of the structure by component type (Chapter 4)
3. Working out support reactions (Chapter 5)
4. Deciding on the manner of application of the load types (Chapter 6)
5. The calculations of stress and deflections (chapter 7)

Some of these steps require supporting theories, methods, and explanations, which will be given as they arise.

Structural data

The first thing is to look at the actual structure in detail in order to find out all the data that we shall need. This consists of geometrical information and material data. What is its size and shape? The dimensions required are not just the length, width, and depth of the component and structure but also the cross-sectional shape. We also need to know the material it is made from because materials have very different properties, such as their strengths (there are many different ones, as we will see later) and their flexibilities – that is, how much they distort under a certain load.

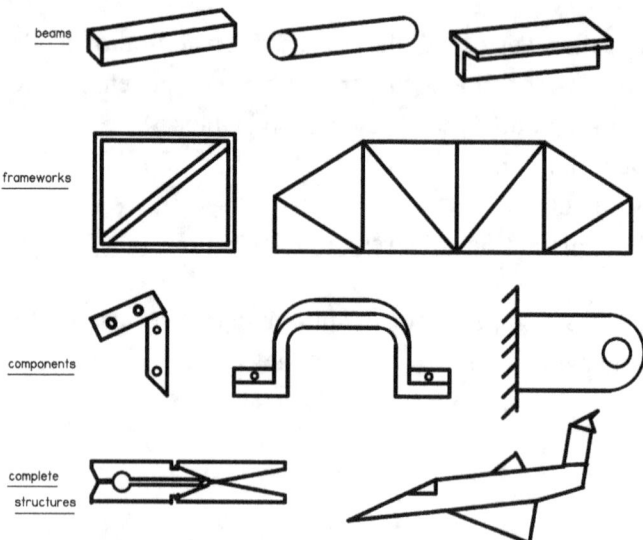

beams

frameworks

components

complete
structures

Figure 3.1 Examples of stress-able structures and components ranging from small individual items to complete assemblies like aircraft

Figure 3.1 shows a few examples of the sort of thing which might have to be stressed. Every possible dimension from each will be needed. In the case of the aircraft, or indeed any complex structure, the data required will fill many books – leave it to the professionals.

Cross-sectional dimensions are dealt with in Chapter 7, Section (a) 9; they are necessary for calculating the stress due to the loading.

Also important will be joint details such as the number and size of the bolts and their material strength as well as the details of welds or soldering.

There are several situations to consider next. First, does the structure exist physically? If so, we are obviously checking it to see whether it is good enough either for its present purpose or for some extended use or a completely different one; or we might be considering modifying it physically. The problem is likely to be that we do not know much about it in the detailed sense: What are the exact dimensions? What material is it made from?

Many of the dimensions can be measured, but some may be more difficult if one side is inaccessible – for instance, the thicknesses of tubes or boxes if their ends are not visible. Indirect methods will be needed; maybe a hole will have to be drilled and plugged later if necessary; perhaps a simple artefact can be weighed and the required thickness calculated. There are ultrasonic instruments that can gauge thicknesses. Photographs can be useful. Imagination in investigation is needed.

You need the technical description of the material if that is possible; otherwise, you will have to make conservative guesses. This could be fatal to your purpose. Steels, for instance, can range in strength by a factor of five! Unfortunately, only if you have access to the original manufacturing drawings is it likely that you will obtain the exact material specification – you probably have to work for the company that made it to get that. You may be lucky and find that it was built to a particular Code of Practice and that will give you much information. There are portable hardness testers which can be used on an artefact, and they will give an idea of a metal's strength, as there is some correlation between hardness and strength. For instance, one approximate equation for steel and aluminium is:

$$Ultimate\ strength\ (MPa) = 3\ to\ 4 \times HB$$

HB is the Brinell hardness. This is one common hardness testing method. There are other hardness testing methods, and you can look up similar equivalences on the Web. It should be emphasized that this is an approximate result, but it will tell you whether you are dealing with a low, medium, or high strength material.

It may be necessary to consult metallurgy experts in their laboratories – they have instruments that can find the constituents of metal alloys. This will give an indication of the alloy and its possible strength range. The hardness will then confirm the type of material.

The material strength of unreinforced plastic components is difficult to ascertain. There are numerous types of plastic, and it is next to impossible for a layman to decide what a particular item is made of just by look or feel, unlike metal. The manufacturing process also affects the strength; the producer should be consulted. In extremis and as a last resort, you may have to cut off a piece of the material and have it tested to destruction in a standard laboratory test.

The strength of reinforced plastic, glass fibre reinforcement (GRP) in most common applications, can also vary considerably. It depends on how much glass has been incorporated, usually between about 30 and 60 per cent, and what form the *lay-up* takes. For instance, at the lowest strength, there is random lay-up where the fibres are an inch or two long and lying in all directions, while the best quality has long fibres which are carefully oriented to line up with the highest loads in the structure. The aircraft and sports industries use carbon as their favoured fibre, as this is both stronger and lighter than glass. However, the fibres themselves can vary considerably in strength, which makes life even more difficult for an engineer trying to assess the strength of an existing structure.

Joints in a structure are very important since they are often the weak points. If the joint is a bolted one you can count the number, measure the size, and, if the markings are visible, read and interpret the material strength from the legend embossed on it. If there are several components joined together in the one bolt group, you have to discover through which each bolt goes.

Welded joints also need examination. Detail weld stressing is quite a complex procedure, and papers and books have been written on it. Look up "weld design" on Google and you will find many more or less helpful guides to design methods. These are often subject to official Standards and Codes of Practice which control the design of potentially dangerous structures such as pressure vessels, cranes, nuclear installations, and so forth. See Chapter 7 (c) (6) for joint strengths. Once you have all the information you can stress your structure.

The second situation to consider is when the structure has not been made at all; it only exists on paper and the drawings and specifications are all to hand. This is easy. You just stress it!

The third situation is when only a general idea of the structure exists. It will be early in the project, when no detail design has taken place. This is the best time to start considering the strength of your structure. The necessary material can be specified, as can the required sizes, and with luck no modifications will need to be done later, as no deficiencies will be discovered. Deficiencies found late tend to be expensive.

It may be that several types of construction have to be proposed, each being stressed. The designs will be strongly, indeed mostly, influenced by non-structural considerations and possibly by the manufacturing capabilities available. Eventually, a design will be chosen and drawings and specifications produced so that you are in the same position as the second situation above.

Summary

Two sorts of data need to be found:

1. Geometrical – detailed sizes of everything as accurately as necessary
2. Materials and their strengths and stiffnesses

In the first situation, the structure exists and you have no constructional information, so you must measure and estimate to obtain the data. In the second, you know all about it, and in the third, everything is yet to be decided.

So all the required information is to hand. Next look at the actual structure.

CHAPTER 4

Classification of the structure by component type

The first thing to do in examining the structure is to decide how to classify its components. Is it made of bars, beams, columns, struts, tubes, plates, shells? Does it have lugs, bolts, welds, rivets? Does it contain cylinders or spheres? Is it made up of boxes or trusses? Most likely, it is a mixture of some of the above.

Examples of beams are the horizontal members of steel frame buildings that you see for a short time as they are being put up. Beams are mostly used to resist bending moments. The verticals in these buildings are known as columns or struts as they carry mostly end loads, although they usually have to resist some bending too. So, we can say that any long, thin, narrow thing is a beam if it resists mostly bending loads; it could be anything from a knitting needle to an aircraft wing, a gate post to a bridge. The length of beams should, in general, be greater than 10 times the cross-sectional depth otherwise a significant part of the load is resisted by shear stress in the beam and not by bending. Beams and so forth can also be curved, a complication which will be discussed later.

Often a long and slender component is called a bar if it resists tension only or mostly, a column or strut if it resists compression mostly. An example of a bar might be a towing bar used to pull along a car which has broken down; it becomes a strut if the towed car is slowed by the towing one and is therefore put into compression. The importance of the compression case is that long and slender components have a tendency to fail by buckling before the stresses are very high, so different calculations have to be done.

Tubes are often used to transfer fluids and may be under pressure. These count as pressure vessels. Alternatively, they could be used as bars or beams, as with scaffolding, for instance. This is because they are lighter – or, more importantly, cheaper – than equivalently strong, stiff, or otherwise suitable solid bars.

Plates are thin expanses of material. They can be any shape, but the only formulae available giving their stresses and deflections apply to rectangular, circular, and elliptical planforms. If the shape under consideration is different, approximations have to be made. A straight-sided but non-rectangular one can be evaluated as an encompassing rectangular plate; an irregularly curved shape might be approximated by a circular or elliptical one, again encompassing for a conservative answer. To do an estimate, it is best to draw the shape and its approximation, one on top of the other, and get a reasonable fit. Then see how high the stresses are – if they are not high, there is no need to get a better value. If they are, consult an expert.

Plates are considered flat; if they are curved, they are called shells. These may be complete cylinders, spheres, or something similar, or they may not. The curvature may be circular or not. If they are, circular loads applied at right angles (or *normal*) to the surface of the shell, such as a pressure, are carried mostly by tension or compression and not bending – that is, they stretch or compress. Flat plates carry such loads by bending.

In flat plates, the thickness should be no more than a quarter of the smallest transverse dimension. It can be much thinner. For pure bending as a load-carrying mechanism, the maximum deflection should be less than about half the plate thickness. If the deflection is much more, then the plate starts to carry the load by stretching; it partly transforms into a shell. Obviously, a judgement has to be made as to how to stress the component. First check the deflection, assuming pure bending. If this is small, then bending it is. If not, then the problem becomes how much load is carried by bending and how much by stretching. This is where things might become difficult. If stresses are low, it does not matter much, unless an accurate stress or deflection is required. If so, or stresses are high, it may be that more sophisticated methods are required and a specialist should be consulted.

Cylinders and spheres and their associated pipes and instruments containing fluids under pressure are a common special case. Again, the relative thickness is important, so if the thickness is greater than about one-tenth of the radius, a different approach to finding the stress is required. This is because the stress through the thickness, in the direction of the radius, becomes important and the stress in the direction of the circumference (the so-called *hoop* stress) varies through the thickness. Practically speaking, a different set of equations applies.

Ropes and braided or twisted wires constitute an interesting class of structure, as they can only take tension. They will not compress or carry a load in bending. The only simple way to find their strength is by testing them. The manufacturers do this, and what you have to do is look up their strength in the catalogue. The other important point about them is their stiffness: how much they stretch under load. Again, this has to found by testing and should be in the catalogue.

Trusses are made up of bars and struts in various arrangements. They can be seen in many types of structures, bridges and electricity pylons being obvious examples. There are simple methods of finding the loads in the individual struts and bars by first considering the overall truss as a single component.

Similarly, box-like structures are made up of plates and shells and may be analysed by considering the total structure as a single component and thus finding the loads in the discrete plates.

This classification tells us what type of analysis to do; in practical terms, what equations to use in order to find the stress or deflection. Although basic physics applies to all the various types of components described here, the mathematics for each situation was solved at different times by different people, so there is a large collection of equations to choose from, each published separately. They may be very simple, just one number divided by another, or they may be very lengthy in certain tightly specified conditions, as we shall see. Some researchers spent ages getting the answer to a particular problem and then publishing their papers. All these results have been collected into various books, as mentioned earlier.

Summary

The structure needs to be defined in terms of the individual components; this will help decide what equations to use to find the stresses and deflections.

CHAPTER 5

Support reactions

Before getting down to the level of individual parts of a structure and finding the stresses in them, the structure has to be considered as a whole or as a single piece. The reason for this is that all the loads on it have to be discovered and the loads supporting the structure are still unknown. The load case under consideration will give the *applied* loads, and these will give rise to supporting *reactions* which have to be found. All the applied loads have to find their way through the structure and be reacted. So if you sit on a stool, your weight (the applied load) has to be reacted by the floor. The stool transmits your weight to the floor. Each load case must be considered individually; the position or direction of the reaction points may be different.

But first we need to think about four matters to help us understand the step. These are as follows:

a) Newton's Laws of Motion
b) Components of a load
c) Determinate and indeterminate structures
d) Constraints

Section (a) Newton and physics

Everyone's heard of Isaac Newton and his Laws and probably also that early twentieth century physicists have proved him wrong. Actually, he was only wrong at each end of the physical world, at the tiny end, in the subatomic sphere and in the cosmological sphere when the velocity of a body nears that of light. In all the ordinary situations that are dealt with in this book Newton is still king. The answers given by his Laws are correct.

They are stated as:

Law 1. A body continues in its state of rest or of moving uniformly in a straight line except insofar as it is made to change that state by external forces.

Law 2. The rate of change of momentum is proportional to the externally applied force and takes place in the direction in which the force acts.

Law 3. To every action there is an equal and opposite reaction.

So there you are in your car, going along at a constant speed, in a dead straight line and on the level. The forces on the car are the thrust from the tyres, which counters the wind's drag exactly and the weight of the car countered by the support of the tyres exactly. If you want to go faster or change direction, you have to apply more force via the accelerator pedal or the steering wheel respectively. This situation describes all three Laws. Law 1 applies to the steady speed state and when changing it. Law 3 is illustrated by the equal and opposite tyre thrust and wind forces horizontally, in addition to the equal and opposite weight and support forces vertically. Law 2 is concerned with changing the speed or direction.

Law 2 mathematically translates into the well-known statement that *force equals the mass of an object multiplied by its acceleration.* When you want more speed, the tyres have to pull harder; when you want to change direction, the tyres have to apply some force sideways.

When something is accelerating and we know the value of the acceleration and the mass of the object, we can calculate the force on the object by using this relationship, which is very useful. Conversely, if we know the force on the object and its mass, we can work out the acceleration, from which, if it's free to move, we can tell its speed and when it will get somewhere of interest, which is occasionally important.

Section (b) Components of a load

Any load can be split into what are called *components*. Loads have two parts – size and direction. The part we are interested in here is direction. Look at the diagram below.

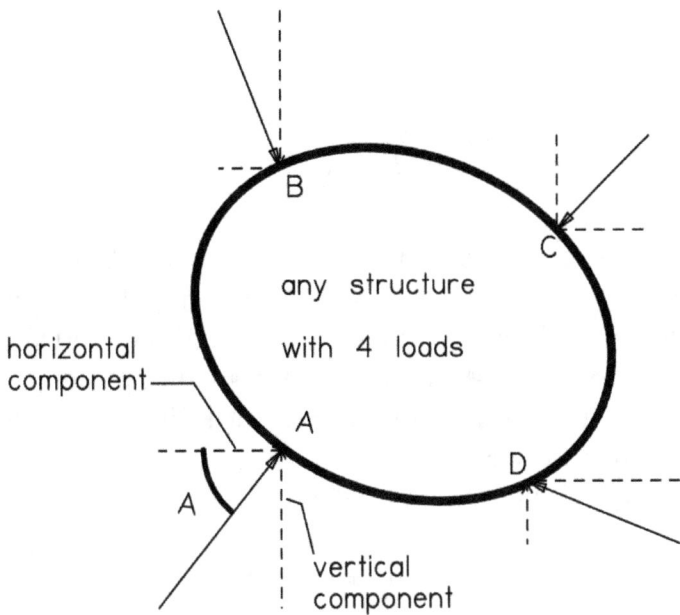

Figure 5.1 The ellipse represents any structure and has four loads applied as shown.

In the diagram, a structure is shown with four loads, at A, B, C, D, applied to its periphery. It is difficult to see whether these loads are in balance even when you know their value because of the angles at which they push the structure. In order to make this easier, we can split each into its vertical and horizontal *components* (not to be confused with components, or parts, of a structure). We use these components of the actual load instead, at the same point of application. The load at A has two components shown, a vertical and a horizontal. These, remember, are instead of the actual load. The others are shown to be split similarly. Then it is simple to add the horizontal components and the vertical components separately. They should balance in each direction if the structure is not to accelerate (that is, the upwards forces equal the

45

downwards and the leftwards ones equal the rightward). A simple bit of maths gives the components.

If the angle of the load at A with the horizontal is A and its value is P_A, then the horizontal component is

$$P_A \times \cos A$$

The vertical component is

$$P_A \times \sin A$$

Sin and *cos* are easily found on most scientific calculators nowadays, so there is no longer any need to look them up in tables, although you can if necessary. This example splits the loads into horizontal and vertical, but we could use any convenient or useful direction as long as the two components are at right angles to each other. This last point is important and will be explained later.

The balance equations are as follows:

$$V_A + V_D = V_B + V_C$$

$$H_A + H_B = H_C + H_D$$

V stands for the vertical component and H for the horizontal component of the load, P; the subscript indicates the location. Each term is given by the expressions involving sin or cos:

$$V_A = P_A \sin A$$

And so on for the other V and H terms.

We must also check that the loads do not have a net turning moment on the structure. So look at every load, or component of every load, find its *moment arm* from some convenient point, and multiply the two together to give you the moment of that force about the point. Add up all the ones tending to turn the structure clockwise and all the ones tending to turn the structure anticlockwise. The two totals should be numerically equal. The structure will then not accelerate rotationally.

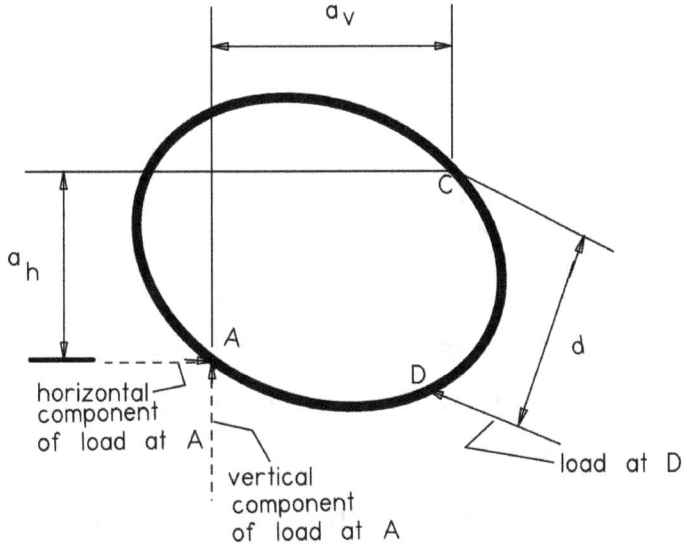

Figure 5.2 Loads and their moment arms

In the diagram above (a partial copy of the previous diagram), there are two loading points, A and D. At A, the horizontal and vertical components of the load are shown; at D, the actual load is shown. The moment arm for each about point C is also shown. Moment arms are obtained by drawing or imagining a line from point C parallel to the line of action of the relevant load and measuring the distance between them. This is shown for each load in the diagram.

The resulting moment balance equation is

$$H_A \times a_h = V_A \times a_v + P_D \times d$$

All three equilibrium calculations above – vertical, horizontal, and rotational – are in only two dimensions for ease of explanation. If the third dimension – into the paper, as it were – exists in the problem (that is, there are loads in this dimension too), you will have to consider it as well. To make it easier to talk and write about the three dimensions, it is usual to call them X, Y, and Z.

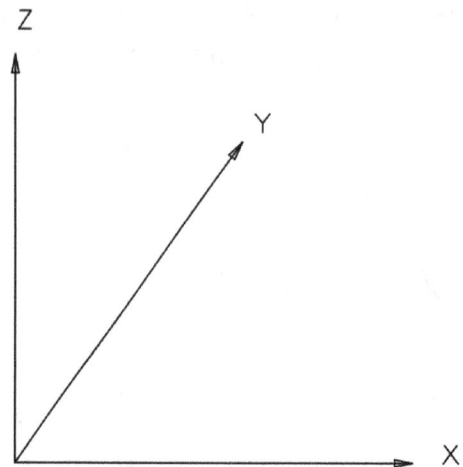

Figure 5.3 Right-hand axes – three axes mutually at right angles

If there are only two dimensions in the problem, they are often labelled X and Y. When the third dimension is relevant, then both the X-Z and the Y-Z plane may also have to be examined. The work is trebled. Remember, the Y-direction is into the paper, away from you and at right angles to both the X-direction and the Z. The direction of the three dimensions is decided by another conventional rule called the Right Hand Rule. The simplest way of visualizing this is to open the thumb and first two fingers of your right hand so that they are at right angles to each other. (Don't strain your hand, though.) The thumb is the X-direction, the first finger is the Y-direction, and the second finger is the Z- direction.

Using this idea of splitting loads into components and using Law 3 gives us what are called *balance* equations, as demonstrated above. There are up to six of them. There are three linear ones in each direction X, Y, Z as in the diagram above and there are a further three rotational ones, one about each of the linear directions. Law 3 says that in each direction, or about each direction the loads must balance. So, in the X-direction the loads to the left equal the loads to the right; the moment clockwise about X equals the moment anticlockwise about X. Loads are for linear actions, moments for rotations.

When the problem can be fully described in two dimensions only three of these equations are required. In the X-Z plane they would be X loads, Z loads and moments about Y. In a three dimensional problem all six may be needed.

Section (c) Determinate and indeterminate structures

Some structures can be solved – that is, the reactions on it and loads in it – by the use of the balance equations only. Structures in such a situation are called *determinate*. Strictly, that should be *statically determinate*, but we need not be that technical. There are two circumstances to consider: the reactions to the applied load's overall structure and the loads within a structure, in individual components. We are, in this part, concerned with the reactions. We will deal with the internal loads later.

So we are looking at a structure and asking how it is supported. All structures have to be supported, be they aeroplanes flying through the air, when the wings do the job; or settled on the ground, when it's the wheels; or a simple stool on its three legs, with the floor doing the supporting. Actually, all loads eventually reach the ground when we are discussing things on this earth. The wings of the aeroplane are supported by the air, which is supported by the ground, as the wheels are when the aircraft is on the ground.

You can identify determinate problems because they need all the structure – without one piece they fall down or collapse. So a stool has to have all three legs to stand up. Again, a beam sticking out from a wall, known as a *cantilever,* with a load dangling from the end needs a vertical support at the wall as well as a rotational one. Remove either and the cantilever collapses, as shown by the diagram below.

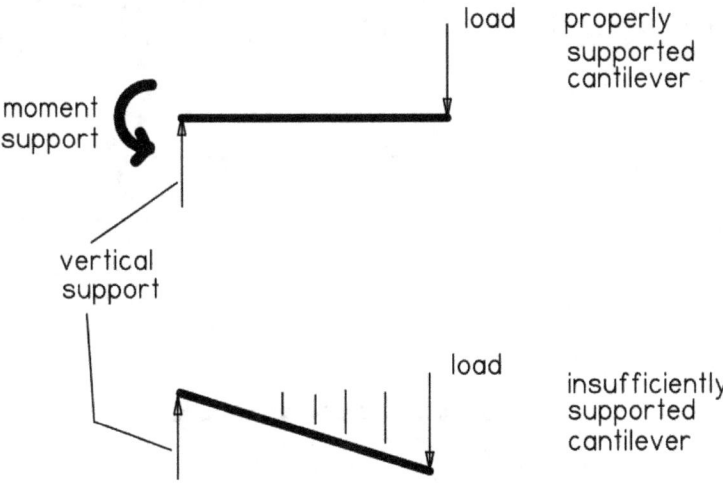

Figure 5.4 Cantilever with and without sufficient support

If, for instance, the moment support is removed, the cantilever would tend to rotate about the vertical support. This shows that the structure is determinate. The vertical support could have been removed instead; the cantilever would then have dropped down without rotation. There still would have been insufficient support, so the structure is determinate. The solution to the question about the size of the vertical and the rotational supports is found using the balance equations. Vertically, the support equals the load; rotationally, the moment equals the load multiplied by the distance between the load and the supported end.

Cantilevers are the simplest of structures, but the principle applies to all structures. If one support is removed and the structure is not then adequately supported so that it moves under the influence of the loads, it is determinate. If the structure just distorts a bit when the support is removed because the loads bend it to a new shape, that is not enough to establish determinacy and it is therefore indeterminate. It must move bodily for the test to establish determinacy. Distortion is due to strains within the structure.

Conversely, indeterminate structures do not need all of the supports. There is a proviso, which is that the structure is strong enough not to need the support which is removed. This has to be assumed true for the purposes of this part of the procedure of solving the structure. Later we will find out whether or not it is. Look at the next diagram of a beam on three supports. The wiggly line represents a *continuous* load – that is, one which is spread all along the beam without interruption (known as a *uniformly distributed load* or UDL).

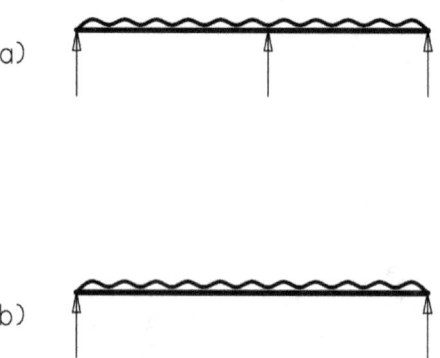

Figure 5.5 Uniformly loaded beam with two or three supports

There are three supports in diagram a) but only two in b). However, the removal of the central support does not necessarily mean that the beam collapses; the two remaining supports can hold it up against the load. Either of the other two supports could have been removed instead; the beam is still held fast. The beam in a) is indeterminate; the beam in b) is determinate. If another support is removed from b) it falls down. There could have been four or more supports – that just increases the degree of indeterminacy. There can be any degree of indeterminacy; advantage can be taken of this indeterminacy by making the structure strong enough to resist the extra stresses of the failure of one of the supports. That occurrence does not then matter. More about this later.

Section (d) Constraints

In practical terms, at our level of analysis, a *constraint* to the support of a component is either *simply supported* or *fixed*. This applies to the support of beams or plates or any similar component. The concept arises out of the mathematics used to find out the deflections and stresses in a component.

The equations that give the deflections and stresses are not quite enough in their initial form; they need a further piece of information. This is provided, in general, by an assumed deflection at a particular point in the structure. The deflection can be either a linear or a rotational one. In the case of the fixed support, it is the *slope* (a rotational deflection) of the deflected beam or plate at the support, and it is taken to be zero. For a simply supported beam, the slope is zero at the centre. Similar considerations apply to plates and shells. We do not need to delve further into this matter here, but it is vital to understand the concept.

The drawing below shows a beam. It is represented by its centroidal line, or its neutral axis. Initially this is straight, but when it is loaded in some way, it deforms under the load to some curved shape. It may not be apparent to the unaided eye, but all beams do this when loaded. Also shown is the slope, which is a tangent to the curved line at the end. The angle is that between the slope and the initial straight line and is a measure of the loading and the stiffness of the beam. The beam is simply supported, and therefore the angle is unaffected by any rotational restraint at the support.

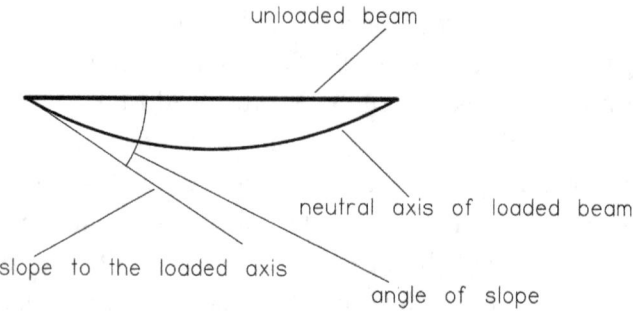

Figure 5.6 Deflections of a loaded beam relative to an unloaded one.

If we now want to choose an equation for the stress or deflection of a part of our structure from one of the lists published by many authors, we must first decide whether the part is fixed or simply supported at its supports. Note that there can be different types of supports at each end of a beam or each edge of a plate. There will be several sets of equations in the lists, one for each combination of *edge conditions.*

Of course, in reality, the slope and therefore the degree of fixity at a support can vary between the fully fixed condition and the simply supported one when it has a value greater than zero. The value of the slope is dependent on the loading and the stiffness of the component. Often it is not easy to decide how fixed a support is, and it is usual for an approximate estimate to assume full fixity (for which the stress can be found), which is safe at the support. It may not be safe at midspan, however, so here it might be wise to assume simple supports. This does not give a consistent answer, as there can only be a single degree of fixity, but it will be safe. If this results in an unacceptable design for some reason, it may be time to consult an expert who may reach for his finite element computer program.

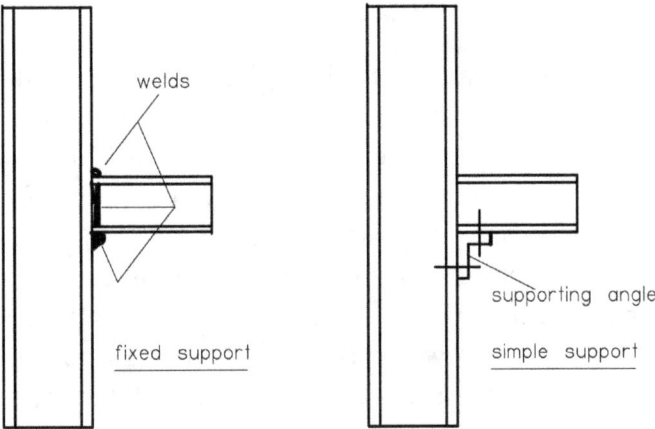

Figure 5.7 Side view of a light beam supported by a much stiffer one. Fully welded gives a fixed support, a bolted angle an SS one.

For instance, if a long, slender horizontal beam is supported by a much deeper vertical one (see diagram above) and it is strongly held by many bolts top and bottom, or welded all round, the assumption of a fixed condition would be pretty accurate – not 100 per cent accurate but near enough. If only a couple of bolts near its bottom edge were all that supported it, it would be better to assume the horizontal beam was simply supported. The effect of this is quite dramatic. The maximum bending stress in the simply supported (SS for short) beam is 50 per cent higher than the stress in the fixed beam when the loading is uniformly distributed along it. For a point load, the SS stress is twice that in the fixed beam.

Every joint in a structure has to be assessed according to this idea. In general, if two parts are strongly joined together by welding (as above) or strongly bolted together or are integrally machined, then they may be considered fully fixed; otherwise, the joint should be assumed simply supported. As indicated above, you could assume both for safety's sake if a little extra material does not concern you.

Section (e) Finding the reactions

We can now work out the reactions with some understanding of the underlying principles and ideas. These are Newton's Laws, the splitting, if necessary, of loads into more conveniently directed components; the establishment of determinacy or not; and the constraints of the joints and supports, whether simply supported or fixed.

So if the structure under consideration (this can be anything from a single beam to a large and complex assembly) turns out to be determinately supported, we can find the value of the supports. We have found the vertical and horizontal components of all the applied loads. We can now apply the balance equations.

By thinking of each of the three directions and three rotations in turn, all the unknown loads can be derived from the known ones. For instance, in the case of a simple three-legged stool with a known weight on it, you would be able to say the total load on the ends of the three legs equalled the weight. If the stool was of symmetrical construction and the load was central, each leg would carry one-third of this. You have taken the vertical balance and solved for the loads applied to the structure without actually putting pen to paper, just by eye.

If the stool is not symmetrical, say one leg was further forward from the back pair, you could *take moments* about a line through the back two legs. The weight times the horizontal distance to this line equals the upward reaction on the forward leg times its horizontal distance to the line. Solve this little equation and you have solved for the leg load using just this Law. If the back two legs are symmetrically placed with respect to each other, then obviously they take half the remaining load each.

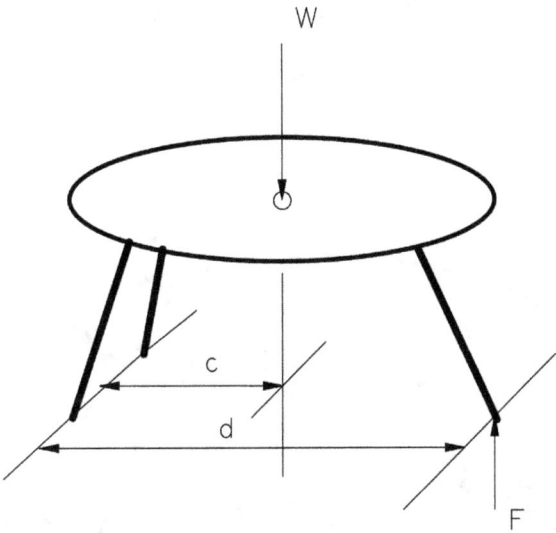

Figure 5.8 Three legged stool with weight W.

Using simple maths, the problem of finding the reactions is solved in the following way:

In the vernacular, take moments about the line through the back two legs:

$$F \times d = W \times c$$

Hence

$$F = W \times c \div d$$

The other two legs share the remaining load equally because the weight is equidistant from each:

The load on each of the other legs $= (W - F)/2$

So the problem was determinate because removing one of the legs causes collapse, and the example demonstrates the use of the balance equations to find the support load in each leg.

Suppose now that there had been another load yet it was a horizontal load on the seat of the stool.

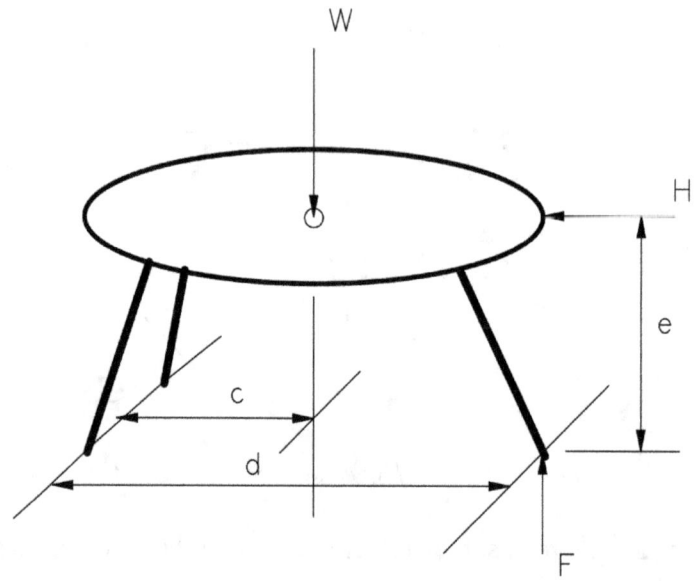

Figure 5.9 Three-legged stool with weight, W, and horizontal load, H

The stool is still determinate. It will fall over if a leg is removed, so the balance equations will solve for the leg support reactions, but the moment has changed, hence these reactions will be altered. The new moment on the stool is the old one plus the effect of the horizontal load.

$$F \times d = W \times c - H \times e$$

Hence:

$$F = W \times c \div d - H \times e \div d$$

The other two legs share the load equally as before:

$$= (W - F) \div 2$$

Two points to notice: First, if the moment due to W is the same as that due to H, F is zero, so there is no reaction on the forward foot. It could be that the moment due to H is greater so the stool tips backwards;

now, unless the foot is screwed down, there is a rotational acceleration on the stool and a further load has to be considered – the inertia of the masses is involved, more of this later.

2) There must now be horizontal forces on the legs at floor level to counter *H*. This will normally be provided by friction. More on this later.

Beams which are determinate are usually either SS at each end – as shown in figure 5.5 (b) – or they are cantilevers (figure 5.4). If there is another support, linear or rotational, they become indeterminate. Rectangular plates which are supported on opposite sides or as cantilevers are also determinate, but they can be treated as beams anyway. Again, another support renders them indeterminate. Cylinders and spheres under pressure are determinate for the pressure loading. This is because they are symmetrical about any diametrical plane, and the balance equations alone will solve for the reactions at that plane; any other loading system makes them indeterminate.

Most structures and components within structures are indeterminate and more than balance equations are needed to find the support reactions. Extra equations come from considering the deflections within a structure – as many as are wanted can be conjured up. *Continuity* is the key technical jargon word by which this process is described, but we won't go into it at all; we leave that to the experts.

Sometimes there are ways of getting around indeterminate structures without heavy mathematics. For instance, consider a four-wheeled trolley which has to carry things along hospital corridors. It will be nice and smooth so small hard tyred wheels will be used; in this load case if we assume that all the wheels touch the floor equally we have an indeterminate structure. If you remove one wheel the trolley does not collapse. Either it stands on the remaining three or it tips a bit and stands on two and the wheel-less corner, depending how the weight is distributed.

The load from a heavy piece of kit may not be central, either from front to back or sideways. It seems that each wheel will have a different load. Which is the largest and how large?

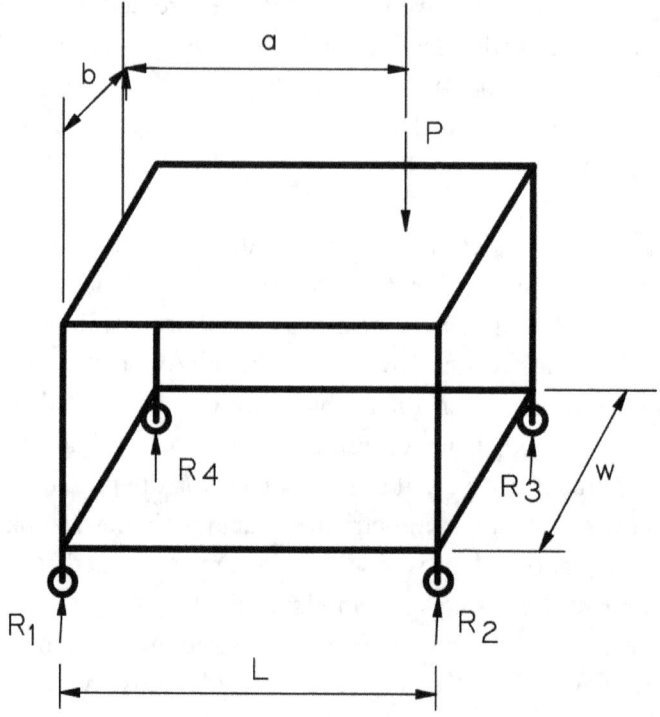

Figure 5.10 Trolley with off-centre load

The reaction at one end to a load which is not centrally placed on a structure is inversely to its distance from that end; the farther the load is from the end, the less the reaction. In the case of the trolley, this means that the reaction at R_2 and R_3 together is

$$R_2 + R_3 = \frac{a}{L} \times P$$

Similarly, the above load can be distributed between R_2 and R_3:

$$R_3 = \frac{b}{w} \times \frac{a}{L} \times P$$

Substitute this into the first equation and you get R_2 and you can do the same sort of thing at the left side of the trolley for R_1 and R_4. Try it. But you do not need to solve this problem.

In fact, no floor is level enough so that it can be guaranteed that all four wheels touch the floor equally all the time. In reality, springs or soft tyres are needed to approach that situation. So the critical load case is when only two diagonally opposite wheels react the applied load with a third just balancing the trolley. In practice, one wheel will take most of the load (as the load is not central) and should be sized for that situation. You may be able to decide exactly how much of the load is reacted by one wheel if you know the maximum possible values of a and b. In this case, R_3 is the most heavily loaded when only R_1 and R_3 are touching the floor.

If the load is as near as possible to R_3, depending on its size and shape, R_3 may take most of the load; so assume it takes all and make the wheel strong enough to cope.

Such considerations often simplify a problem sufficiently to get a required answer without solving the structure in detail.

If the trolley is relatively heavy, its weight should be added to wheel reactions. If it is doubly symmetrical, then each wheel reacts about a quarter of the trolley's weight in the softly sprung case, fairly obviously; otherwise, it reacts half of the self-weight of the trolley.

Circular arches and cylinders under pressure, which is a uniformly distributed load, present another interesting determinate case.

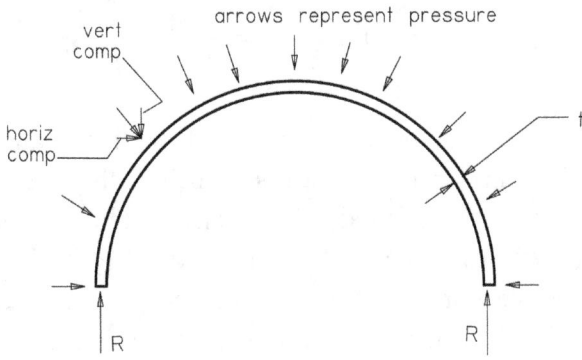

Figure 5.11 Semi-circular arch under pressure

The first thing to notice is that the horizontal component of the load due to the pressure at any point has another equal but opposing one on the opposite side of the arch; the total horizontal load on the arch is zero, although the arch will distort a little under the loads. The vertical component of the pressure load is balanced by the reactions at the supports, R. These are equal; this can be seen immediately due to the symmetry of the structure and you can prove it by taking moments about the point of application of one of them. This will be the same as taking moments about the other.

To get a load in Newtons from a pressure (N/mm^2), you have to multiply by an area. We will take a unit depth (into the paper), a millimetre in this case. For the other direction, we realise that the total horizontal length is the projected length of the arc (2r). This is the total of all the horizontal length projections on which each element of pressure acts.

Taking the radius of the arch to be r, the pressure p, and the reactions R, the vertical balance equation gives

$$2R = 2r \times p \times 1$$

$$R = pr$$

This is true for every unit depth of arch (into the paper). The stress is then obtained by merely dividing by the thickness, t:

$$stress = \frac{R}{t} = \frac{pr}{t}$$

This is often known as the *hoop* stress in the cylinder, and it is twice the *longitudinal* or *meridional* stress, which is the stress along the cylinder. This fact can be found by balancing the total load on the lid of the cylinder by the longitudinal load in the wall. Try it.

This example puts the cylinder into compression; an internal pressure, obviously, causes tension in the wall. It is essential to stress such cylinders, as failure can be dangerous due to the large amount of energy stored in fluids which are at high pressure.

A number of frequently occurring examples giving reactions are listed in various books, as has been mentioned before. The examples are single components such as beams, columns, rings, arches, plates, shells, cylinders, spheres, and so forth, both determinate and indeterminate. The books do much of the work for you. You do not have to set up the balance equations, where it's possible, as above; you have to substitute into equations instead. Typically, the reactions are one of the results given.

Most are for indeterminate situations, but they do have limitations because they usually are for uniform material properties and cross-sectional shapes and sizes. Mostly they describe single span beams and so forth. When equations for multi-spans with different sections are given, they are very long and therefore error inducing. Much care is required.

For instance, take the case of what is known as a *propped cantilever*, which is a cantilever (determinate) with an extra support at the other end, which makes it indeterminate. See the diagram below.

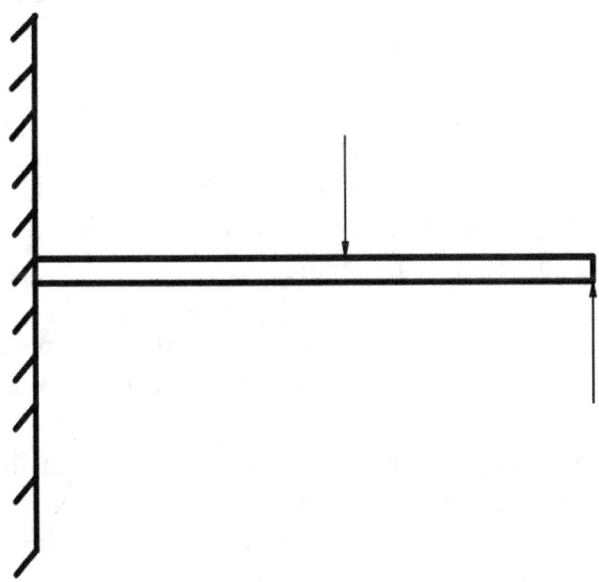

Figure 5.12 Propped cantilever with a central load

The rake-like hatching at the left end represents a fixed support and implies a support moment and a vertical or horizontal supports as necessary. The prop is the right end support. The equations given by the books, e.g. Roark, are:

$$\text{Fixed support reaction} = \frac{11}{16} \times \text{load}$$

$$\text{Prop support reaction} = \frac{5}{16} \times \text{load}$$

$$\text{Fixed moment} = \frac{3}{16} \times \text{load} \times \text{length of beam}$$

Roark's also gives the distribution of moments and so forth along the beam and equations for the deflection. This is probably the most comprehensive collection of equations for analysing simple structures. The old fourth edition is the easiest to use. Later ones, which have additional authors, become both more extensive and more complex.

Section (f) Summary

This chapter has been concerned with finding the reaction loads on a structure for every load case, which must be done before we investigate the stresses and deflections in the individual parts. Four concepts and ideas help in achieving this.

1. Newton's Laws: These provide the fundamental physics for the subject. There are three laws, all of which we use at one time or another.
2. Components of a load: The idea that the direction of action of any load can be usefully split into two directions (at right angles to each other), generally though not necessarily vertical and horizontal. Together with the Third Law this leads to the balance equations.
3. Determinate and indeterminate structures: Determinate structures can have their supporting reactions solved by using only the balance equations. Indeterminate ones need more equations or techniques. This concept also applies to individual components within a structure.

4. Constraints: At the level of this book, a constraint is either simply supported or fixed. These apply to how a structure or component is held by its support. In practice, there can be constraints which are intermediate between these two conditions.

CHAPTER 6

Loads and the manner of their application

There are many types of loads and many ways in which they apply to a structure. Most of the load types have been considered in the chapter on derivation of loads – for example, mass, pressure, and so on. Now we will consider exactly how to deal with the manner in which they exert their force on a structure. They are classified under a number of headings; we will consider them one by one.

These loads can be either static, which do not varying with time, or dynamic which do vary with time. This is not necessarily the same as a moving load. A moving load can be represented as a static one – a so-called *quasi-static* case. The pressure of fluids due to moving something, a boat through water or wind over a sail, can be applied to the structures as if there were no motion, just a pressure. Similarly, when something rotates relatively quickly and constantly so that centrifugal forces are generated, these can be applied as if there were no rotation. In such situations, the load, which does take into account the velocity of the structure, represents the motion adequately.

Section (a) Point loads and moments

Point loads have been mentioned before in the examples. They are a mathematically theoretical concept. You cannot actually have a point contact between two objects and transmit any force, as the contact point immediately gives way and the load is spread over an area. The size of the area is dependent on the hardness of the material and the size of the force. So-called Hertzian stresses are developed under this patch of material – they are mostly, not always, not damaging and are ignored unless the contact patch is very small.

Point loads are a useful concept for circumstances such as the leg of the stool in one of the previous examples because a very short distance from the area of contact, this simplification does not matter when the stresses in the local structure are being calculated. The manner of support for the load has no influence on the stresses in the rest of the structure. So if we are interested in the stresses at the top of the leg, for instance, the precise details at the bottom where it meets the floor do not matter. We need only the position of the centre of the contact patch of foot and floor.

When a load has been classified as a point load, its effect on a structure can then be found from the published books of equations under the point load classification.

There are many because of the various end conditions and the position of the load along the beam.

Similarly, moments are assumed to be applied at a point. At the point of application, there obviously has to be some structure, perhaps the attachment of a cantilever to its support, as in Figure 5.7. The local material around the attachment will have a complex stress pattern, but at a distance less than the depth of the cantilever, this will disappear or have a much-reduced size and the stresses due to the overall loading on the structure will prevail.

Again, the assumption that the moment is applied at a point is a useful one for the stressing of the rest of the structure, and there are equations in the books to enable you to do this easily.

Section (b) Line loads

These can be thought of as many point loads very close together in a line along a beam or across a plate. They are also known as continuous loads, as mentioned before, when describing Figure 5.5. The mathematics used to obtain the equations is a little different so there are a complete set for line loads. There are different types of line loads in that they may not be constant from one end of the line to the other. The example in Figure 5.5 shows a uniformly distributed load – that is, it does not vary. Obviously, the variation could be of any profile, but those that have been analysed are usually confined to a triangular shape; this declines linearly from a maximum to zero. It need not start at one end of a beam and finish at the other; it may load the beam only partially. The various loading profiles can be added to make up an approximation of any actually occurring one.

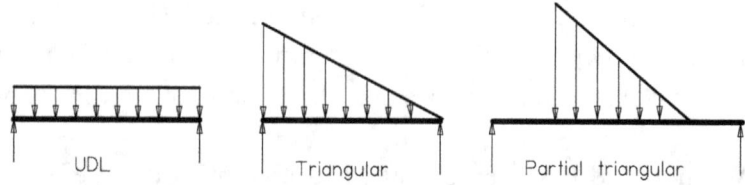

Figure 6.1 Beams with various types of distributed loading

It is possible to have similar line moments, although these do not occur so often.

Section (c) Torque and Couples

Torque or twisting force, which has the same units (force x distance – e.g. N.m.) as moment, is usually to be found winding up bars or tubes. There are a different set of equations to describe the stresses and circular deflection for this type of force (see Chapter 7 (c) (4). In this case, the torque acts about the centre line of the bar or beam. See Figure 6.2 (a) below. If there is a pair of equal point loads acting along a beam, in opposing directions but displaced in point of application, they are said to apply a *couple* to the beam. If the displacement is along the beam, it is like a point moment outside the couple. See Figure 6.2 (b) below. The beam, though, is in bending.

If the couple is applied through subsidiary structure across the beam, it will twist it. See below in Figure 6.2 (c) below. This is one way the torque, T, in Figure 6.2 (a), could be applied to the beam.

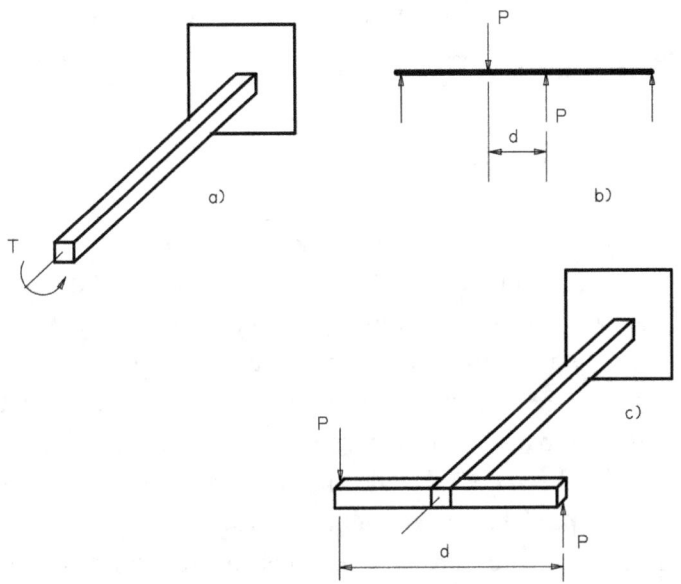

Figure 6.2 a) Torque, T, twisting a cantilever; b) a couple,
$P \times d$, *on a beam; c) twisting a beam through subsidiary structure,*
again $P \times d$.

Section (d) Pressure

As has been explained in the chapter on the derivation of load cases, pressure acts all over a surface and normal to it. Fluid pressure is usually, in static cases, constant over the surface between the seals which stop it from leaking away. This is not so in the case of a wall restraining a fluid such as water. The pressure increases linearly with depth, so the wall experiences a higher pressure at the bottom than at the surface of the water, where, of course, it is zero. The wall has triangular loading, as in Figure 6.1, except that this should be rotated through ninety degrees. For example, those garden swimming pools which are not sunk into the ground have a pressure at the bottom acting on the walls of one-tenth of a bar, or one-hundredth of an MPa, for every metre of water depth.

In the case that the fluid is moving, the pressure when it flows past a curved surface is not constant over that surface. The variation from the static pressure may or may not be significant in comparison with static pressures in a particular case. This dynamic pressure is calculated as

$$\text{Dynamic pressure} = \frac{1}{2} \times \rho \times v^2 \times C$$

Where ρ is the density of the fluid in kilograms per metre^3, and v is the velocity of the fluid in metres per second. C is a constant describing the shape of body in the fluid stream; it is found experimentally, and there are reference books giving many examples. This gives the answer in Newtons per metre^2. This is a small number if we are concerned with slow moving air, but it increases with the square of the wind speed, as anyone who has been out in a gale knows. Moving water, because of its higher density, produces much higher forces.

An interesting pressure case is when a pressure vessel is being tested. If the pressure in the test is anywhere near the failure pressure of the test piece, the test must take place in a special enclosure. The latter must be designed to cater for the pressure and for any flying debris. An empirical result from tests of pressure vessels to failure is that about 65 per cent of the energy in the pressurized fluid is expended as blast and 35 per cent is absorbed into the kinetic energy of the debris of the broken vessel. If the vessel merely splits and there are no free-flying missiles, then that energy is absorbed by distortion of the walls of the vessel, the shock wave, and so on.

The pressure loading is in two parts, a shock wave which lasts about a millisecond and a longer-lasting lower value pressure which quickly builds to a maximum and decays slowly. The actual value of these pressures depends on the circumstances of the test piece and the enclosure. It is best to design pressure test enclosures with one wall or the roof open to the atmosphere so that when a failure occurs, the blast is vented to the atmosphere quickly, with little hindrance. The response of the structure is dependent on its natural frequencies, as well as the energy in the pressurized fluid, and is complex. The stress analysis is best left to experts.

The 35 per cent of the energy that is contained in any flying pieces can be analysed because of an interesting finding. This is that the velocity of all the pieces is initially the same, whatever the size of the piece. The energy is kinetic energy which is calculated from the well-known equation

$$KE = \frac{1}{2}mv^2$$

Here, m is the mass of the vessel and v the initial velocity of the bits. If the energy of the pressurization is known (it equals the pressure times the mass of fluid), this equals the KE and the initial velocity can be calculated. Using the equation again on any piece whose mass is known gives the KE of that piece. The walls of the enclosure must then absorb this energy. This mechanism will depend on the wall's construction and again is complex, needing the attention of an expert.

In general, it is best to make pressure vessels of a material that is not brittle, as this will minimise the number of pieces in any failure. Indeed, if the material is ductile enough, such as low- or medium-strength steel, there may be no flying debris, just large splits which can release the pressurized fluid. Tests on pressure vessels should be performed with liquids rather than gases if possible. Compressed liquids, being nearly incompressible, do not expand much before losing their energy whereas gases, highly compressible, expand much more.

Pressure loading also exists in non-fluid situations – for instance, between a tyre and the ground, a flat-bottomed plate and a table, and in many other situations.

Section (e) Body forces

Some forces act on every molecule of a structure and are known as *body forces*. They usually arise from accelerations acting on the body and can be thought of as three-dimensional pressures. Gravity is the most common example. They are calculated from Newton's Second Law:

$$P = m \times a$$

P is the force in Newtons, m the mass in kg, and a the accelerations in m/sec². In the case of gravity, it is $a = 9.81$ m/sec². This can often be simplified to 10 so as to make calculations easy.

Section (f) Slope-induced components

When a surface or beam or whatever supports a load which is at an angle to the line of action of the load, there are usually two components to the load which have to be considered: one along the surface or beam and one at right angles to it. The effect of the slope or angle between the load and structure is to produce two forces.

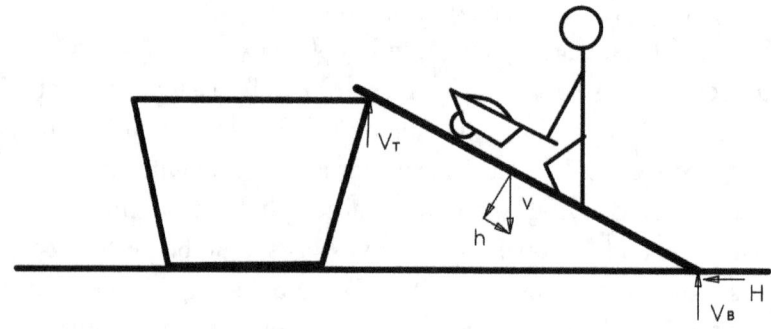

Figure 6.3 Wheelbarrow supported vertically, V, at each end and horizontally at road level

For example, think of a wheelbarrow being pushed up a plank which is leaning on a skip so that its load can be dumped into the skip. The weight is vertical, and each end of the plank has to be supported vertically. But also, the plank will have a tendency to slide backwards, which has to be resisted by friction between it and the road. This effect is calculated in a similar way as taking components.

To work this out, you start with the mass of the man and wheelbarrow at their joint centre of gravity (CG) and realise that this acts vertically. But you want the load on the plank at right angles to it; this enables you to choose its thickness. To go with this component is one which is directed along the plank, and this is the slope-induced component.

It is applied to the plank by friction under the man's foot and is reacted by forces H and V_B at the road; H is the friction generated by the plank on the road; h is the sine of the angle through which the plank is lifted multiplied by v, if you do the geometry.

Another similar example is the potentially dangerous one of a ladder, where the foot must be stopped from sliding outwards.

This effect occurs whenever a load is applied at other than zero or ninety degrees to a beam or other member. It may be applied through another beam joining it or directly. The load may be a point load or a continuous one.

Section (g) Pre-tension

The stresses in this type of load are due not to physical loads such as wind or sacks of stuff or whatever but to forced distortions. This can occur in many ways; one of them is due to temperature changes, and this will be dealt with separately. The most common is probably due to tightening a bolt. That bolt may then be put in tension by another load.

The reason a bolt is tightened well beyond finger tightness is to make sure it does not easily shake loose. The mechanism by which this works is friction. Friction is dependent on how hard the two items concerned are pushed together. To slide a book across a table takes a certain force. If five more books are placed on top of it, then the sliding force is increased proportionately to the increase in weight. Similarly, if the bolt is done up tightly, the friction under the head and the nut and between the threads increases. This friction works against any undoing tendencies. Incidentally, in many situations, this is not good enough and further locking methods must be employed.

Finding the stress in tightened bolts under external loads is a little more complicated because the effect of the *abutment,* those components squashed by initial tightening of the bolt, has to be taken into account. This is done by considering the deflections in the bolt and abutment. Look at the diagram.

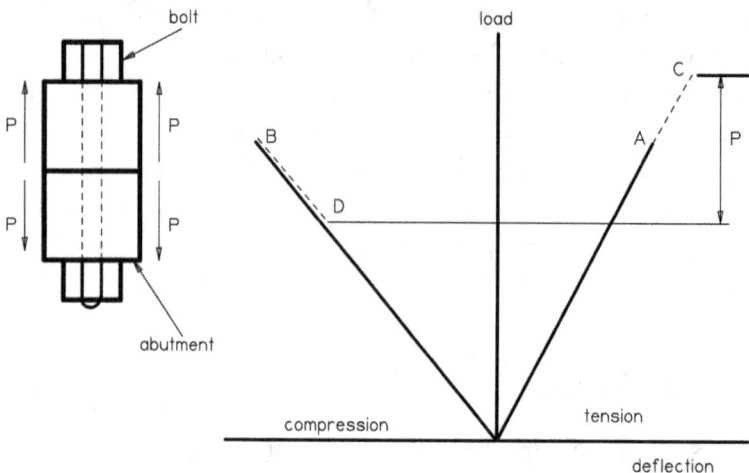

Figure 6.4 Load in a pre-tensioned bolt

First look at the left drawing. This represents a nut and bolt holding together two components. Initially the load, P, is not present, but the nut is tightened on to the bolt with a spanner so that it is very tight. This obviously stretches the bolt and also compresses the abutment – that is, the components. Then the load is applied in some unspecified way. This is represented by the graph on the right of the figure. Initially, the bolt is stretched into tension up to A, while the abutment is compressed to B.

The line to A and C represents the load/deflection of the bolt, and its slope is a function of the stiffness of the bolt (in turn a function of its size and material). So any point on the line tells you the stretch you get in the bolt for a given load. Similarly, the line to B and D represents the stiffness of the abutment.

The tension load and the compression load are equal. When the load P is applied, the bolt stretches farther to C. Because the bolt lengthens, the abutment can also lengthen, or relax, thus losing some of the compression. Its deflection reduces to D.

If the balance equation is applied to the structure, it says that the bolt load (at C) equals the applied load (from C down to D) plus the compression in the abutment (from D to zero). The main point to notice is that the increase in load in the bolt as the load P is applied is less than P. It is sometimes stated that when a load is applied to a properly pre-tensioned bolt, there is no increase in the bolt load. This, it can be seen, is not true. The increase in bolt load is merely reduced, an important reduction, as will be seen when fatigue failures are discussed.

When the compression in the abutment reaches zero, if the bolt has not broken, all the applied load is resisted by the bolt and the abutment is not compressed at all. This may, however, lead to *fretting,* or scraping between the faces of the components, which is not good.

Theoretically, therefore, we can calculate the load in the bolt from the graph. However, we have to know the properties of the materials – that is easy – and the size of the bolt, length and diameter, also easy, but in addition the area of cross section of the abutment which is compressed. This last is often difficult to estimate. If the abutment is a cylinder not much bigger than the bolt head or the diameter of a good thick washer, then it is fair enough to assume that the entire abutment is equally compressed. If the abutment is much bigger than the bolt head, then the compressive stress in it will diminish as the distance from the bolt hole increases. If a precise answer is required, then an expert practitioner will have to use his Finite Element computer program or some tests will have to be conducted. As a quick and crude approximation, take the area compressed to be equal to a cylinder whose outside diameter is twice the bolt diameter and see whether you have a low stress solution.

This information can now be used to calculate the bolt load. There are two parts, the initial load due to tightening the nut and bolt and the extra load in the bolt due to the applied load.

When tightening a nut and bolt, the spanner load or torque has to overcome the friction between the underside of the nut and its mating face as well as between the meshing faces of the threads. The friction is dependent on the materials of the nut and the washer or component onto which it bears and dependent on the lubrication or its lack between all the faces. There is a wide variation in the friction (five to one) between un-lubricated and well-lubricated threads and faces, so it is necessary to be aware of this. However, the application of the tightening torque does pull the bolt into tension as the threads slide over each other. The question is how much tension a given torque will cause in the bolt. By the way, there are special spanners (so-called torque spanners) which will apply measured values of torque to nuts or bolt heads.

For average conditions, which would be new nuts and bolts with light lubrication, there is a rule of thumb which gives a fair idea of the relationship between the torque and the resulting load in the bolt. For a better estimate, consult specialist literature such as ESDU 86014. This gives the torque/tension relationships for a wide range of tests results which have been collected and analysed.

The rule of thumb:

$$P = 5 \times \frac{T}{D}$$

P is the induced tension load in the bolt, T is the applied torque and D is the bolt diameter.

From the geometry of the graph sketched above, it can be calculated that the extra bolt load, beyond the above load (often called the pre-tension load), is given by

$$P_D = \frac{P_L}{1 + \frac{A_a E_a}{A_b E_b}}$$

P_D is the extra bolt load due to the applied load, P_L is the applied load, A_a is the area of cross section of the abutment, E_a is the Young's modulus of the abutment material; similarly, A_b is the area of cross section of the bolt, E_b the Young's modulus of the bolt.

Examination of this equation in a couple of instances is interesting. Take the case of steel nuts and bolts clamping together two steel components. This means that the Young's modulus of each is the same and cancels; also assume, as suggested, that a cylinder twice the bolt radius is compressed by the pre-tension.

$$A_a = (2 \times r)^2 \times \pi - \pi \times r^2 = 3 \times \pi \times r^2$$

$$A_b = \pi \times r^2$$

$$\frac{A_a}{A_b} = 3$$

$$P_D = \frac{P_L}{1+3} = \frac{P_L}{4}$$

This means that the extra load in the bolt due to the applied load is only a quarter of the applied load.

If we have aluminium components, the Young's moduli are not equal. Aluminium has only one-third the modulus of steel.

$$\frac{A_a \times E_a}{A_b \times E_b} = 3 \times \frac{1}{3} = 1$$

$$P_D = \frac{P_L}{1+1} = \frac{P_L}{2}$$

The extra load is now half the applied load, twice what it was with the steel components. I do not know of a text book explaining this – this explanation is based on what we used to do at Rolls-Royce.

ESDU is a commercial organisation which was born of the old Royal Aeronautical Society data sheets which used to find out and measure useful engineering facts for the aircraft industry when Britain had a dozen or more aircraft companies. Their products are rather expensive for individuals but can be accessed in large or university libraries.

There are other situations in which structures are prestressed at the manufacturing stage and where a similar principle of analysis can be used. Often pins are forced into undersize holes in order to fix and hold them tightly. Yachts have their masts held up by wire ropes which are screwed down tightly so that they are in tension and the mast is put into compression; also, the hull structure is bent between the mast foot and the shroud fixtures. Prestressed concrete beams have their reinforcing bars on one side of the beam pulled tight before they are anchored. This increases the load the beam can carry. The outer surfaces of components are sometimes shot-blasted (metal or ceramic particles are literally fired at the surface); this puts an outer layer into compression, and when subsequently the component is pulled into tension, this outer layer suffers a lower tensile stress – good for fatigue resistance. This is a specialized procedure.

Section (h) Temperature-induced forces

As described previously, if a structure or a component is assembled or manufactured at one temperature and the temperature then changes, it is likely that some strain, and therefore stress, is induced by the change. This can be because the structure's expansion or contraction is restrained by another structure; or because it is made of different materials which have different rates of expansion (that is, different so-called *thermal expansion coefficients*); or because the temperature varies across the structure.

Except in a few simple circumstances, as in the example above, it is difficult to calculate thermal stresses by hand. Nowadays, computer techniques called Finite Elements are used, which will be mentioned later. It may be possible to approximate a complex situation by making simplifying assumptions provided they overestimate the stresses. This will give a safe structure. The process of safe simplification is an important subject which will be gone into later; it provides a quick method of stressing all structures which do not need to minimize the material they use. The manner of application of these thermal loads depends on the shape of the structure itself.

Thermal strains can be *transient*, as when a structure heats up, or *steady state*, when the temperatures reach their final value. A number of simple examples of stress calculations can be found in various papers collected in the aforementioned book by Roark.

Section (i) Magnetic and electrical forces

It is obvious that magnets exert forces; it may also be important to remember that electrical conductors with currents flowing in them generate magnetic forces. To find these forces, refer to standard electrical engineering textbooks. They will apply their force to the structure as point loads, line loads, or torques, according to how these force agents are connected to it.

Section (j) Summary

This Chapter describes how loads are classified and applied. Nine common types of loads are identified. There will be more, more obscure or specialized types.

CHAPTER 7

Calculations

Before actually calculating stresses and strains, there are some more matters to understand. All calculations are supported by a number of underpinning physical and mathematical laws. These define the boundaries of the applicability and the accuracy of the procedure. Sometimes equations giving stresses and deflections are used where they should not be, because the underlying conditions of the structure and the underlying assumptions supporting the equations do not match. A simple example is given by a beam which is simply supported where an equation assuming fixed end supports is used. The structural assumptions are wrong for the problem. Again, for mathematical reasons, only if a structure behaves linearly can stresses due to several loads be added to obtain a total stress. So it is essential to establish the correct conditions governing the structure and then to choose the matching mathematical equations.

Section (a) Mathematics, physics, and useful techniques

There are at least *14* points to consider and keep in mind when embarking on a stressing procedure. Some we have mentioned before, some are just useful techniques, some are bits of mathematical knowledge that help to arrive at a correct(ish) answer, and there is a little more physics.

1. Newton's Laws

These have been given and explained in Chapter 5, section (a). They are the basis of the physics of stressing, as has been said before.

They should always be borne in mind, nothing can contradict them, so reference to them is a useful check at any time in a procedure. In Chapter 5 they were used to produce the idea of balance equations so that the support reactions on a complete structure could be found. These balance equations can be used on any part of a structure to check that intermediate results are correct.

2. Linearity

There is another early scientist who had a direct effect on stressing – Robert Hooke (1635–1703). He found that the deflection of a structure was, up to a point, proportional to the load on it. If the load was doubled, the deflection doubled. This was so for complete structures and its individual components, large or small.

This is true because of the nature of many materials. The stress in them is also proportional to the strain (strain is a function of the deflection) up to a certain level, called the limit of proportionality. This will be explained in Chapter 8. After this level, the proportionality breaks down.

This effect is called *linearity* because if you plot on graph paper the deflection against the load, you get a straight line. It has useful implications for stressing. Firstly, obviously, the deflections and stresses of a structure are proportional to the load. So if you hang a 100 kg load on something, you will get certain results which you (having studied this book) can calculate, even if this takes some time and effort. If you now decide to hang a 200 kg load on the same place, you need not spend the same amount of time and effort doing the calculations all over again. All you do is double the original results. So long as the resulting stresses do not exceed the limit of proportionality for the material, they are correct. Obviously, if the new load is 137 kg, you multiply by 1.37 (1.37 = 137/100). The only exception to this rule is if any component is in compression; the buckling stress may be less than the limit of proportionality. You can use proportionality only up to this point.

Secondly, you can add the results of two or more completely different load cases, again provided the total stresses do not exceed the limits. Each load case is solved separately, and all the results can be added together; obviously, this includes the support reactions as well as the stresses and deflections.

If the structure's behaviour to loads is not linear, then neither of the above techniques can be applied. A structure or component behaves in a non-linear fashion due to a number of conditions. Firstly, the material might be non-linear – plastics, for instance, or a metal which is already stressed to the *yield point* (this term will be explained in the chapter on materials). Secondly, if a slender component is in compression, it is only linear up to the point at which it buckles. Thirdly, if the support conditions change during the loading or gaps appear in a structure, then it must be judged to be non-linear. If a graph of load on a structure against deflection is not a single straight line, it is non-linear.

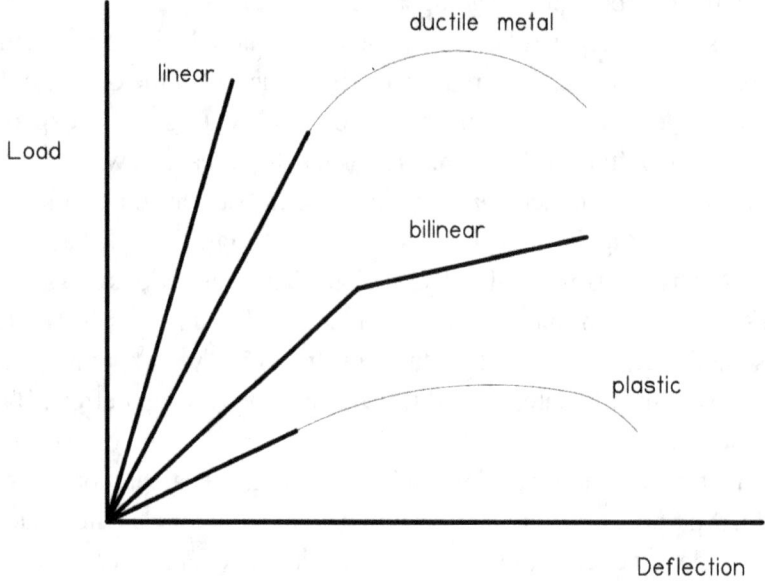

Figure 7.1 Load – deflection curves for various material types

In the graph above, the single straight line is the usual load-deflection condition assumed in stressing. The line marked *ductile metal* displays the behaviour of a metallic structure and a component; it is linear up to a point. The *bilinear* line is two straight lines and represents a structure where the support conditions change when a certain load is passed. Unreinforced plastics are continuously non-linear and are represented by a curved line, while reinforced plastics are controlled by their reinforcing material which is mostly linear.

3. Slow load application

It is necessary to apply loads to structures slowly. This is to avoid having to deal with the energy involved in movement. Theoretically, loads are supposed to be applied infinitesimally slowly, which is of course impossible. Quite slow is good enough, but with anything too quick, extra care must be taken. For instance, if a mass is placed carefully onto a beam and slowly lowered until the beam stops moving, a certain deflection and stress is imposed on the beam; this is known as the *static* case. But if the mass is held just touching the undeflected beam and then released suddenly, the deflection and therefore the stress nearly doubles, although only for an instant. It eventually settles down with the same deflection as the static case. If the mass is dropped from a greater height, then the peak or maximum stress will be greater still. The beam must be made of a sufficiently large size to cope with the greatest deflection and stress, so this is an important point.

If loads are applied slowly enough to avoid any over-deflection – in other words, if the loads are supported by other means until the structure in question develops its full deflection before being completely released – then the condition is fulfilled and static stressing will be accurate. For instance, if a crane lowers a crate onto a lorry and the crane wire does not slacken completely until the lorry is supporting the crate, then static conditions apply. If the crane operator slips the clutch when the load is only just touching the lorry deck or even when it is still clear of the deck, then dynamic conditions apply. The load now has some distance to fall and develops *kinetic* energy which must be absorbed in the structures. This is more complicated. A specialist ought to be consulted if such a situation applies in your case.

4. Stationary conditions

The balance equations and Newton's Laws demand that the structure being stressed is either stationary or in a steady (that is, non-accelerating) state. Actually, in many instances, the condition can be fulfilled even if the structure is moving or accelerating, provided that loads are applied which represent the acceleration or some effect of motion – say, air pressure due to a moving structure. Then the structure can be assumed to be stationary, but the effect of the motion or acceleration is recognized as a set of loads. This type of situation will have to be identified when embarking on a stressing task.

5. Load components

This subject has been covered in Chapter 5, section (b). It will be necessary to decide which, if any, of the loads have to be resolved into convenient directions.

6. Load circuit

This point has been used in Chapter 5, which considered support reactions. According to Newton's Laws action and reaction are equal and opposite. This implies that a load has to go somewhere; it has to be met, as it were. At the overall structure level, there have to be supports, as explained in Chapter 5. Within the structure, this is also true; if there is a source of load built into it, such as a hydraulic jack, then the loads have to be absorbed. They can be tracked around the structure and must meet up and cancel each other. For instance, consider the diagram of a lifting bridge.

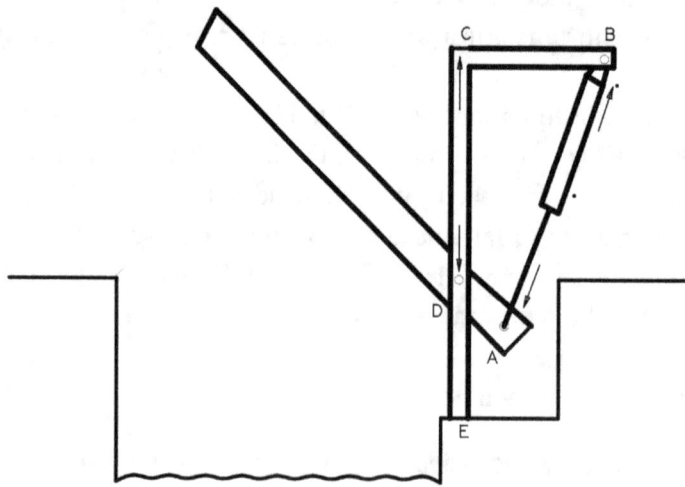

Figure 7.2 Diagram of the loads in a lifting bridge

When the jack is pumped up, it exerts forces at each end, A and B in the diagram. The vertical component of the force at B travels to C, bending the beam BC as it goes. The force then stretches the post CD to the pivot at D. The force next travels to A via the piece of bridge structure, where it cancels the other jack force – a complete circuit. The same sort of thing happens with the horizontal component. Work it out for yourself. The whole bridge is, of course, supported at E, and this support load is not affected by the jack force.

It is necessary to visualize these load circuits for all loads when stressing structures. Some loads – for instance, the load at E, which is due to the weight of the bridge – are said to be earthed; no further consideration is given as to where they go next. You do, of course, have to have proper foundations, but that is another story.

There is an analogy here with electrical circuits. No current can flow unless there is a complete electrically conducting route for the electrons to travel around. Again, the earth can be part of this. The analogy extends to the mathematics, where, for instance, the voltage in a circuit is analogous to the force in a structure. Before digital computers, problems in complex structures could be solved by building an analogue electrical circuit.

In the diagram, the structure is simple and the load paths are obvious. Often, though, the situation is more obscure because there may be several possible load paths. If there had been a strut between B and D, then some of the load from the end of the jack would go, in tension, from B to D. How much? This depends on the relative stiffness of the two paths. Another calculation would be required. Since tension members are stiffer than bending ones, it is probable that most of the load goes down the strut and little in bending BC. For initial calculations, assume this to be the case. Obviously, this is true only if the members are all much the same size; also, the connections must be sound. It is no good having undersized or slack bolts.

7. Partial structures

This concept is very useful when stressing a component part of a large, more complex structure. This might be one bar in a framework or a panel in the structure of an aircraft or a beam in some multi-beam structure. The idea is to replace all the rest of the structure by the load it exerts on the component of interest.

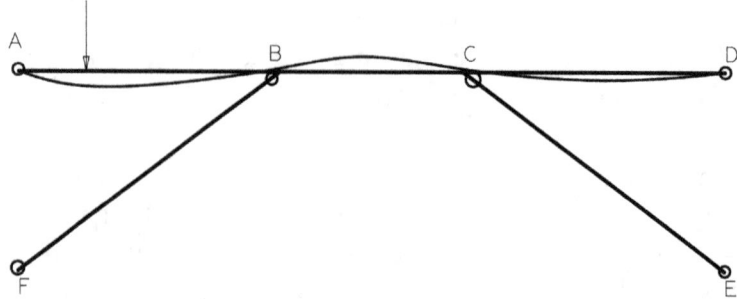

Figure 7.3 Continuous beam, ABCD, with two central supports

Figure 7.3 shows a long beam, ABCD, simply supported at its ends and propped at B and C by bars FB and EC with *pin joints*. They are called pin joints because they allow the beams to swivel at those points, and therefore no moment can be applied to the bars. The beam, ABCD, is continuous, however, so moments generated in AB can be reacted by the remainder of the beam. There is a load, P, on the beam AB and some of it is reacted at D, E, and F as well as A. The structure is indeterminate because the beam is continuous and supported at two intermediate points. The sense of the bending deflections, much exaggerated, is shown in thin lines. This is a useful thing to do with all structural problems, as will be explained in the next section.

If we jump ahead and assume that the structure has been solved for the loads at the joints, we can decide on the necessary strength of beam BC. To do this, we ignore the other beams and consider BC in isolation – but with the joint loads. Note that beams are assumed to carry transverse, to their centre line, loads – that is, *shear loads* and moments – so only these will be considered initially to keep things simple. The next diagram shows the situation.

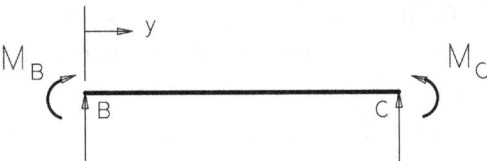

Figure 7.4 Beam BC extracted from Figure 7.3 with end loads representing the rest of the structure.

If the loads are known, then the beam BC can be stressed no matter what the rest of the structure looks like. The bending moment at any point, y millimetres along the beam from B, is M_B plus the support load at B times y. The shear is equal to the support load. There is also an end load which arises because of the slope of the two supporting bars, EC and FB. Often in stressing beams, the stresses due to such end loads are small and can be ignored, but care must be taken that this is true.

Provided a sufficient number of the loads are known, this procedure can be used on any structure, which is useful for sizing new components or just checking that existing ones are strong enough.

8. Visualizing deflections

Under load, all structures deflect. This deflection can be and often is so small that it is invisible to the eye, but it is always there. The magnitude of the deflection depends on the load, the size of the structure, and the stiffness of the material. Steel deflects much less than unreinforced plastic, other things being equal. It is very useful to be able to imagine these deflections, as in Figure 7.3. The deflections drawn there are much exaggerated. You can easily see that the load pushes span AB downwards into a curve. Since the support at B is simple and the beam is continuous, this curve must be continued through B. It is now heading upwards, but the support at C forces the beam down again. Similarly, the support at D forces another reversal of the slope of the curve to D. The supports at B and C actually move upwards a little due to the stretch of the supporting bars which is not shown.

This picture shows where the maximum bending takes place, at the highest curvature midspan (but don't forget the bending at the supports), and where the stress is tension and where it is compression. If you do calculations to find the actual deflections and they come out in the opposite direction to that shown, you should suspect your calculations. In this way, you can quickly gain a qualitative understanding of how your structure works, where the high stresses are, where a large deflection may cause interference, and so on.

9. Properties of plane areas or things you must know about your components

So far, we have looked at components as we normally see them in use. Now we must look inside them, as it were. For beams, this means the cross section which is what you see if you were to cut a beam at right angles to its length. It is obvious that the fatter a beam or bar is, or the thicker a plate, the stronger it will be, other things being equal. For a simple tension load, this is exactly true. The area of the cross section is all that matters. But for bending, the disposition of the area is also important, as it is for torsion. For simple, or pure, compression, the disposition of the area is important after a certain point – that is, the buckling load.

The area is easy to calculate. Width multiplied by breadth equals area of a rectangle. If the cross section is made up of several rectangles, do each separately and add them up. Triangles have an area equal to half the base times the height. School stuff!
However, it turns out that for situations other than pure tension, where the area is in a cross section is important.

When a beam is bent so that it curves, say downwards, the top becomes compressed and the bottom stretched. This implies that the middle is neither, so there is no point in having much material in the middle since it is not needed to resist the bending. In fact, the area is wanted at the top and bottom, which is why the beams one sees in half-erected buildings have cross sections like an **I** or an **H** on its side. They are often called **I**-beams. The area is at the extremes, with a thin connecting web. The area, or rather the material it represents, exists only where it is worked hardest.

A bar in compression behaves like one in tension – that is, the area is all that matters, until its buckling point, and then it starts to bend so it needs to be shaped like a bending beam. Bars in torsion need their material near the outside of the bar, so hollow bars are nearly as good at resisting torsion as solid ones. Obviously, they are lighter.

The geometrical function of the cross section required for bending is called the *second moment of area*. It is also known as the inertia (incorrectly, as inertia should also involve mass) or just as the I of a section. To find it for a particular section, it is easiest to look it up in a suitable book. Standard steel sections are given in booklets published by the Steel Federation, and they will send them for a modest fee. Many reference books have them listed, including *Machinery's Handbook*, which can be found in any reference library. *Roark's* has a list as well as a simplified formula which applies to compact, solid sections and gives the I without too much error. The other way, for complicated sections, is to calculate the I (see addendum to this section). This usually occurs when a beam is made up of a number of differently shaped pieces of material or has a unique shape.

Another important result from this source is the position of the neutral axis, which is where the deflection (compression or stretch – see above) is zero. If the cross section is symmetrical about its centre, then that is also the neutral axis position. If it is not, then the displacement from the centre needs to be calculated.

For the record, the second moment of area is the sum of all the elements of area times the square of their distance from the neutral axis. This is an important property of any bending member.

The only transverse property of a plate in these sorts of calculations is its thickness.

10. Stiffness

This is a useful concept when thinking about load paths – that is, how much of a load travels down each of several possible paths. It is made up of geometrical and material characteristics. Consider a beam of uniform (unchanging) section with SS supports at its ends and with a centrally placed point load. It is reacted equally at each end, so the load travels to the left and to the right equally because the stiffness of the left portion of the beam is the same as that of the right portion. One of the constituents of the stiffness of a beam is its length, in this case the length of the path from the load to each support. If the load had been much nearer the left end, then a larger portion of the load would have gone to the left support. The left part of the beam is stiffer because it's shorter.

If, with a central load, the left half of the beam had had a larger I (second moment of area – see previous section), then more of the load would have been reacted at the left support because a larger I makes the beam stiffer.

Also, the material of the beam is important. Intrinsically, every material has a measurable entity called its Young's modulus (usually called E), which has been mentioned before. Steel has a high one, other common metals have lesser ones, reinforced plastics are adjustable according to their make-up, and unreinforced plastics have very low ones.

So stiffness is proportional to I and to E but inversely proportional to length in beams. Other structures are similar in principle but in the case of tension load paths, it is the area of cross section, not the I that matters. In plate components, this reduces to the thickness. Provided you can see all the structure involved in supporting some load, it should be possible to guess where the load goes. Bear in mind that tension is much stiffer than bending for similar sized components.

If you wish to make a better estimate, you must do some calculations. Essentially, you have to take each load path separately and calculate the deflection at the load application point relative to a mutual fixed point, which may be Earth. Obviously, this deflection has to be the same for each; otherwise, the structure bursts apart. If you choose the correct fraction of the load going down each path, this equality can be arranged. An equation is required with the fraction as the unknown. This, by the way, is the basis of all advanced structural analyses.

A similar situation exists for structures whose largest stresses are shear stresses. A panel in plane shear is much stiffer than a member in torsion. So stiff structures are in tension (or compression up to the buckling stress), and shear, flexible are ones in bending and torsion.

It must be remembered that stiffness also depends on the size of the members, so hefty beams in bending could be much stiffer than a light strut. This has to be calculated. If the situation is too complicated, reach for an expert and his computer.

Access panels usually do not constitute a good load path, especially if there is another structure that can carry the load instead. If a panel or door is bolted down, then an unknown factor enters the equation, and that is friction. A tightly bolted joint will transmit load transversely (that is, in shear) to the bolt by friction. The bolt does not need to touch the sides of the bolt hole. However, a shock to the structure, perhaps a hammer blow or an earthquake or something, will break the friction and the structure will become slightly misshapen, with possibly unacceptable consequences. Vibration will have a similar effect. In addition, you cannot be sure that the bolts are as tight as necessary. If they are loose, they carry no load through friction at all; if half tight, only half the load. Doors and panels are best seen as non-structural items as far as the main structure is concerned. However, they may carry load themselves, such as pressure. That load is then transferred to the main structure, of course. Consider that situation separately.

As another example, consider two beams of equal length fixed to each other at their centres and crossing at right angles. They are supported at their ends (that is, at four places) and loaded by a point load at the crossing point (see the figure).

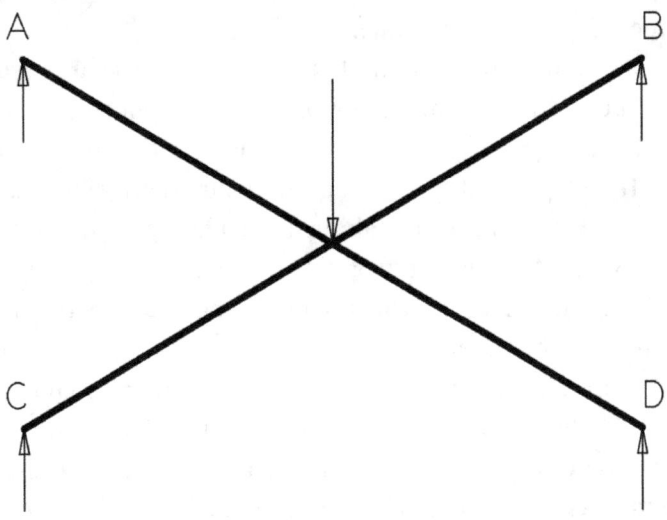

Figure 7.5 Crossed beams with central load

If they are of the same material and identical cross section, then fairly obviously (or strictly by symmetry) the load will divide into four equal parts and each will be reacted at a support. If, however, the I of one of the beams – say, BC – was only half that of AD then the reactions at B and C would carry less of the load and more would go to A and D. The way that this distribution of load is found is by calculating the deflections of each beam. These two deflections must be the same; otherwise, the structure has broken. *Roark's*, for instance, provides a formula for the deflection of a beam with a central load. If you feel strong, try it for yourself. Decide that a fraction, f, of the load P (that is, fP) is carried by AD and therefore (1-f)P is carried by BC and use the formula to calculate the central deflections. Equate them and solve the equation for f. Note that the E, and the length are the same and so cancel in the equation. This gives, as you might have expected, that twice the load is carried by the stiffer beam because it has twice the I. So it takes two-thirds, while the other takes one-third.

This is relatively easy because there is only one factor in the equation, which is different for each beam (the I). All the factors could be different and the same equation would be used, but the arithmetic would be longer. More care is then required. The figure shows that both beams are simply supported but some of the supports could be fixed. If this were so, then the equation to be used would be different; however, the procedure would be the same.

Another point to bear in mind is that the supports are assumed to be rigid. But one or more might not be. If B were not able to carry one quarter of the load and gave way a bit, the situation would get more difficult. If it supported none, A and D would carry half each. C is inactive, which you can see by taking moments about the centre of BC, looking towards A. This is a worst-case scenario, and it might be all you need to know for your purposes, but if you need a better answer, it may be time for the experts.

Notice also that the structure is redundant. One arm could break and the load could be carried by the others, a useful safety feature.

If now there were a vertical post under the load and in line with it, a new fifth load path exists. The post would carry some of the load in compression. Its stiffness is due to pure compression, a much stiffer type of load capacity, and if its dimensions are similar to the bending beams and if its support is good enough, it will carry most of the load.

11. Mathematics

The serious mathematics – that is, solving the differential equations underlying the structural problems and the manipulations of the equations – can be left to the mathematicians. Their answers, in the form of simple equations for deflections and stresses in a structure, can be treated as being in the nature of a black box. You substitute numerical values for the symbols in the equation, do a bit of arithmetic (using a calculator if necessary), and the answer comes out. You do not need to know what the mathematician knows; it is nice if you do, as you will have a deeper understanding, but you can do a useful job without that. A professional stressman would not ever want to spend time solving a basic differential equation in the course of stressing a structure. He would not have time; too many people would be waiting for the answer. Most equations that can be solved have been solved by now anyway.

This, however, does not mean that there aren't a few things to understand. Already it has been pointed out that you must examine the structure and its components to decide on the support conditions, simply supported or fixed, so that the right equations can be chosen.

You should be aware that the equation of your choice may be approximate. If they develop a theory that is very complicated, mathematicians have to make simplifications in order to get a solution. In simple terms, they look at their equation and cross bits out. This is all right if the bit is not significant, so that it produces only a small error in the final result. Alternatively, the simplification may produce a limitation on the applicability of the result. You need to know. Approximations may be introduced merely because of the mathematical method.

If the equation of choice is based on empirical results – say, a series of tests – you have to decide whether your case has the same conditions. If so, you can use the equation. Deviations which are apparently quite small in the conditions can produce surprisingly large errors. You should never use an equation without knowing that its provenance is good.

12. Errors

In the main, there are two classes of error, systematic and particular. In the former, a mistake is made that affects all subsequent work. The data is wrong or the wrong assumptions are made, leading to the wrong choice of equations used to solve the structural problem. It is no good looking through the calculations to find this sort of error if it is suspected that the answer is wrong, for they are in fact right according to the assumptions and data. The only thing to do is review the assumptions and data continuously. If you can, compare the present situation with similar previous ones that have been proved to be accurate.

Particular errors are of the type when you add two to three and make six of it. The problem is that you tend to learn this sort of error, and if you merely redo the calculation in exactly the same sequence, you will make the same mistake. So if you are adding a column of numbers, the thing to do is to add from top to bottom the first time and bottom to top the second. Obviously, the answers should be the same. If using a calculator, that error will not occur. Instead, you are in danger of pushing the wrong key. The same technique works to check this method as well.

This sort of mistake occurs because people, all people, lose concentration regularly. Apparently, the US army did an experiment. Taking alert young students as subjects, they found that they lost concentration every twenty or twenty-five minutes, even if only momentarily. Unfortunately, in calculations, you can make a mistake in a very short moment. When designing Concorde, the supersonic airliner, we had a saying in the office: if the calculation was more than half a page long, it was wrong the first time you did it. There were many thousands of pages. The consequence was that everything was checked and double-checked. In a series of similar calculations, it is worth plotting the answers on a graph. Often one point stands out of line and is therefore suspect. If a computer is used to do repetitive calculations, it is important to ensure the data is correct.

The best checks are independent of the basic method. You may be able to check the answer with a problem which has a measured result or there may be a different way of finding the answer. Getting to an answer in two different, independent ways is best. These answers will not necessarily be identical because of the differences in the method, so a calculation is subject to assumptions in the method and an experimental result is subject to experimental inaccuracies. But if the resulting differences are small in relation to the answer, then it can be assumed that they cross-check.

Anyway, do not accept an answer the second the calculation is finished. Always at least think about it, compare it with previous similar situations, do it again differently, or try to compare it to a practical, measured result. Although in the last case you have to make sure it is a valid comparison in all particulars. If you work out a stress from a complicated bending moment, say a beam with many different loads on it, you can estimate an average load and do a quick calculation assuming a simply supported or a fixed-ended beam. The answer should not be wildly out.

As a tip, if you find you have to evaluate a long equation, and there are some very long ones, do it in pieces. For instance, this is the support moment at one of the ends for a fixed-ended beam with a partial triangularly distributed load:

$$M = \frac{W}{l}\left(\frac{d^3}{l} + \frac{c^2}{18} + \frac{51c^3}{810l} - \frac{c^2b}{6l} - 2d^2 + dl\right)$$

W is the total load, l the overall length of the beam, and b, c, and d give the position on the beam and length of the distributed load. Unless you are very confident and used to this sort of thing, you should not launch into this by putting the numbers straight into your calculator. First, write out the equation above in symbolic terms. Second, write it out substituting the values for the symbols, using plenty of space on the paper. Third, most lengthily, calculate each power – the squares and cubes – and write out the whole lot again with these numbers. Next, calculate each of the six terms separately, writing out each number. If the lines of writing are set out so that every term has its subsequent values under itself, it is easier to check the work. Finally, sum the terms in the bracket and multiply by the term outside. This takes several lines of working and may seem tedious, but it is more likely to yield a correct answer. There are computer programs in which you can write out the equation symbolically, give the values of each symbol, and the program gives the answer at once. Magic. Again, you now only have to ensure that the input data is accurate.

13. Significant figures

The question arises as to how many significant figures the calculations should be done. Significant figures, of course, do not include trailing zeros in large numbers or leading zeros after a decimal point in small numbers. It's the non-zero numbers that count (sorry). Embedded zeros do matter, obviously.

In general stress work, it is rare to have to work to better than 1 per cent. A few Newtons per millimetre squared in a few hundred usually do not matter. This implies that three significant figures are enough. The mathematicians tell us to use one more figure to protect the accuracy of the last important figure, so that is four significant figures. There are exceptions.

One of them occurs when two nearly equal numbers of uncertain value are subtracted. Due to some mathematical quirk, the uncertainty in the result will be worse than in the original numbers. Consider the numbers 100 and 90 and assume they have a 1 per cent uncertainty. The 100 could be anything from 99 to 101, and the 90 could be 89 to 91, to two significant figures for simplicity. Let us lay out the sum as follows:

$$101 - 89 = 12$$
$$100 - 90 = 10$$
$$99 - 91 = 8$$

The first line shows the calculation when the 100 is at its largest and the 90 at its smallest, while the third line shows the reverse. The nominal sum is done in the second line. You would normally do this one without worrying about the accuracy of the numbers. You will notice that the answer has an accuracy of 20 per cent (2 in 10), far worse than the 1 per cent accuracy of the original numbers; actually, the uncertainty of 20 per cent is the worst case; it is up to 20 per cent. This is one example of what is common to any work with uncertain numbers. It gets even worse when the numbers are closer and when the uncertainty is larger.

The situation arises when two load cases occur at the same time and they give stresses at some point in the structure that are opposite in sign – that is, one is tension and another compression. One has to be subtracted from the other. The more accurate the numbers are, the better. This leads to another point: do not, if possible, use careless, broad estimates of loads and so forth. Do not think of a number and add a bit for luck. You end up not knowing where you are. It is better to be accurate all the time. As suggested, this is mostly using four figures, as much as you can in some situations. If you are a bit wary of your result, use a bigger safety factor.

This applies also to rounding numbers up or down. Do not do it except as the last thing in your calculation. As you can see from the above example, severe errors can be introduced if you are not as disciplined as possible.

14. Manipulation of equations

If you are familiar with this process, you can skip this bit. If not, this section is to show how to get a little more out of the equations you choose to use. Most equations in the books give the stress in or deflection of a component, but you may know the stress your material is allowed. You want to know what size the component has to be to achieve just that stress.

As an example, take the equation for the stress in a fixed edge square plate at the middle of its edge when it has a uniform pressure over its entire surface.

$$\sigma = \frac{0.308 \times w \times a^2}{t^2}$$

Above, w is the pressure, a the length of the square's side, and t the thickness. You want to know how thick it has to be. The old-fashioned rule for moving a symbol or number from one side of the equals sign to the other in a simple equation in which there is only multiplication and division is that it moves diagonally. So the σ swaps places with the t^2.

$$t^2 = \frac{0.308 \times w \times a^2}{\sigma}$$

What you are doing, of course, is dividing both sides of the equation by σ and multiplying by t^2, then cancelling.

If you have to choose from standard thickness plates and do not want to pay to machine a thicker one down to your calculated size, obviously you choose the size which is greater than you need.

Another example, in which there are two terms on the right-hand side, is the expression for the buckling load of a straight bar, fixed at one end, free at the other, under both an end load and a uniformly distributed load along its length.

$$P' = \frac{\pi^2 \times E \times I}{4 \times L^2} - 0.3 \times p \times L$$

P' is the buckling or critical load, E is Young's modulus, I the second moment of area, L the length of the bar, and p the uniformly distributed load. Of course, π is the well-known ratio of the circumference of a circle to its diameter. This is a good approximate formula. What is wanted is, say, the I. So move the second term on the right-hand side across the equals sign and the π^2 and E diagonally down, while the $4L^2$ moves diagonally up:

$$I = \frac{4L^2}{\pi^2 E}(P' + 0.3pL)$$

Note the brackets. Both terms within must be multiplied by the one outside and also that, as is more usual, the multiplication signs are left out. From this I value, you can choose the dimensions of the bar.

15. Summary

This Chapter attempts to describe some of the background to strength calculating procedures. It also gives a few of the tricks of the trade, the shortcuts and simplifications. It should raise awareness of the things to be borne in mind when analysing a structure.

16. Addendum – how to calculate the I of a composite section

Consider the cross section of a typical stiffener for the skin of a structure such as an aircraft fuselage or wing or some structural steel for a building. There are many instances where similar sections are utilized. This is shown in Figure 7.6 below. The I required is that about the centroid, or centre of area, of the section, which is known as the neutral axis, or NA, as mentioned before. The position of the NA is also important for the calculation of the stress at any point in the cross section.

Formally, I is the sum of the element areas times each element's distance from the NA squared.

$$I = \Sigma A(c - y)^2$$

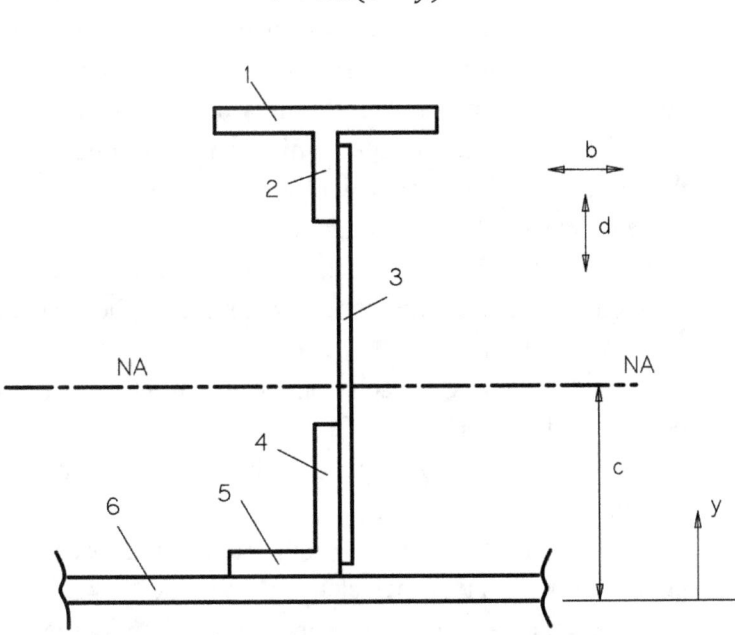

Figure 7.6 Typical composite section laid out for section analysis

The technique to find these has been arranged as a table as shown, which is explained in the notes following it. The justification for the technique requires some maths and is not given here.

Item	Width, b	Height, d	y	A = bd	Ay	Ay²	$I_s = bd^3/12$
1	20	2	38	40	1520	57760	-
2	2	9	33.5	18	603	20200	122
3	1	33	19.5	33	644	12550	2995
4	2	10	9	20	180	1620	167
5	10	2	3	20	60	180	-
6	20	2	1	40	40	40	-
Sum, Σ				171	3047	92350	3284

Notes

1. The idea is to break the section into rectangles as shown. It may be necessary to use triangles or even curved-sided elements; as long as you know each element's centroid, it does not matter.
2. Number the elements and note their width and height. Take the skin width as 20 x its thickness; this is an approximation, as not all the skin will necessarily work with the bending section due to shear lag (which we won't go into).
3. Calculate the position of each element's centroid, y, from a convenient base line, which in this example has been chosen as the outside skin line.
4. Calculate the area, A, of each element as width times height and then calculate Ay and Ay².
5. Sum these last three columns. The Greek letter Σ is often used to mean "the sum of".
6. The last column is a correction to the method because it is based on the assumption that each element is compact. The I of a long narrow element about an axis through its centroid and at right angles to its length is bd³/12; this is added for each element that

sticks up very far. Don't bother for the others. It's worth about 8 per cent in this case, as you will see.

7. First we calculate the position of the NA. This is the sum of the Ay's divided by the total area.

$$c = \frac{\Sigma Ay}{\Sigma A} = \frac{3047}{171} = 17.82$$

This is nearly halfway up the section height of 39, which looks reasonable, as the outside angle has more area.

8. Having found the NA at c from the outside, we can use it to move the I from the axis at the outside skin to the NA. A theorem called the parallel axis theorem can be applied. We calculate the total area multiplied by c^2 and subtract this from the I so far, which is $\Sigma Ay^2 + I_s$.

$$Ac^2 = 171 \times 17.82^2 = 54300$$

9. So the number we want, $I_{NA,}$ is given by the following:

$$I_{NA} = \Sigma Ay^2 + I_s - Ac^2 = 92350 + 3280 - 54300 = 41330$$

If 52190 had been near 94730, then a subtraction error could have been caused, as explained before. To avoid this, it would be better to choose the baseline not at the outside skin but, say, halfway up the section; c would then be small. However, some of the values of y would be negative, which could be overlooked or cause addition errors in that column. It is also more awkward to calculate the y's. If the dimensions are in mm, then I is in mm⁴.

We can find the stress at any point up the section by using the bending equation:

$$stress = M\frac{y}{I}$$

M is the moment at the section; y is the distance from the NA, not from the outside edge, as in the diagram above, which can be calculated from the diagram; and I is I_{NA}.

Section (b) Units

In most of the world, the basic system of units is the SI system (Systeme Internationale d'Unités). However, many structures and components predate the establishment of this hopefully one day universal system; furthermore, the United States does not necessarily use it.

The basic units we need for structural analysis are length in metres, mass in kilograms, time in seconds and degrees of temperature, either Celsius or Kelvin (absolute). These temperature degrees are equal in size but are zero at different points, the freezing point of water or the lowest possible temperature, respectively. In stress work, mostly millimetres are used so that stress is in N/mm^2.

This usually gives stress values in the hundreds, which are judged to be more useable. Too many leading or trailing zeros are clumsy and lead to errors.

Other much-used units are defined in terms of these. The table overleaf lists all those which are relevant to simple structural analysis. They are mostly SI – the pressure unit, the bar, is the obvious exception, as it is from the CGS (centimetre, gram, second) system. There are many others, of course, but they mostly apply to other disciplines. They only interact with structural analysis in more difficult situations, the most common of which is the thermal one. This book will not cover thermal calculations.

Obviously, it is necessary to have consistent units in order to communicate with other people and make sense to them. If the system of choice is the SI one, then we talk of stress in MPa, lengths in metres, and so forth, and everyone understands what we mean. If, incidentally, you start a conversation in MPa, say, there is an implication that you are using SI and you must say whether something else is not in SI. If you use a mixture of systems, say tons per square centimetre, and others think you are using SI, misunderstandings will occur. This has happened – a space mission failed because of a mix-up in units!

Since it is all too easy to mix systems, you could easily have information about a structural situation in a variety of units, so you will need to convert them to a uniform system. You can usefully write your equation and include the units:

$$\sigma = 5000 \, N \div 0.75 \, inch^2 \dots\dots\dots\dots\dots\dots\dots\dots\dots (a)$$

This immediately indicates that you need to change the inches to millimetres for a consistent set of SI units. Furthermore, you can easily see how to use the conversion factor correctly. For instance,

$$1 \, inch = 25.4 \, mm$$

or

$$\frac{1 \, inch}{25.4 \, mm} = 1 \dots\dots\dots\dots\dots\dots\dots\dots\dots\dots (b)$$

An equation, of course, means that everything, when calculated out, on one side of the equals sign has the same value as everything on the other side when that is calculated out. When you multiply or divide anything by one, you do not change its value, so if you multiply the RHS of equation (a), above, by the LHS of equation (b), which equals one, you are not changing its value. You get the following:

$$\sigma = \frac{5000N}{0.75 \, in^2} \times \frac{(1 \, in)^2}{(25.4 \, mm)^2} = 10.3 \, N/mm^2 = 10.3 MPa$$

You need to square equation (b) so that you get in². Of course, 1² is the same as 1, so you can still multiply an equation by that and not change its value.

Now, notice that the in² on the top line of the right-hand side of the equation cancels, as it were, the in² on the bottom line, leaving you with N/mm², which is what you want. It is not strictly mathematically correct to use the term cancel as the unit's name, in^2, are not numbers. But this technique, of cancelling units at the top and bottom of an algebraic expression, always works and is especially useful if you have a long equation.

Here are two tables which might be useful. The first gives most of the SI units used in stressing, while the second gives some conversion factors between SI and imperial units.

SI units

Quantity	Name	Symbol	Definition
length	metre	m	fundamental
mass	kilogram	kg	fundamental
time	second	s	fundamental
temperature	degree, Celsius	°C	fundamental
area		A	m^2
volume		v	m^3
second moment of area		I	m^4
force	Newton	N	$kg.m/s^2$
line force			N/mm
stress	Pascal	Pa more usefully MPa	N/m^2 N/mm^2
pressure	Pascal	Pa more usefully MPa	N/m^2 N/mm^2
fluid pressure	bar	bar	10 bar = 1 N/mm^2
strain		ε	m/m
frequency	hertz	Hz	1/s
energy	joule	J	N.m
plane angle	radian	rad	2π radians = 360°

Conversion Factors

Quantity	multiply	by	divide	Quick approximate guide
length				
	m	3.281	ft	
	m	39.37	in	
	m	1000	mm	
	ft	304.8	mm	
Mass	kg	2.205	lb	
	tonne	1000	kg	
	ton	1.016	tonne	
	ton	2240	lb	
Force				
	lbf	4.448	Newton	
	tonf	9964	Newton	10000N or 10 kN
	kN	1000	Newton	
Stress				
	MPa	10^6	Pa	
	MPa	145.0	lbf/in^2	
	tonf/in^2	15.44	MPa	
Fluid pressure				
	MPa	10	bar	
	bar	9.807	depth of fresh water, m	1 bar = 10 m approx.
	MPa	98.07	depth of fresh water, m	1 MPa or N/mm^2 = 100m
Density				
	lb/ft^3	16.02	kg/m^3	
	lb/in^3	27675	kg/m^3	

The table operates forwards and backwards – for example, multiply "m" by 3.281 to get "ft", divide "ft" by 3.281 to get "m".   Remember that MPa equals N/mm² numerically.

Section (c) Stress and strain

Stress takes into account the load on a piece of structure, the manner of application of the load, and the structure's general shape. If you imagine a piece of material to be made up of many small fibres bundled together, then conceptually stress is the load on the smallest *fibre* of the piece you can imagine divided by the cross-sectional area of the fibre. Stress is a structure's reaction to a given load or force, and different structures will react differently to the same load. Similarly, if the force on a given structure is changed, then the stress will change; so if the shopping bag you are carrying becomes heavier, the stress in your arm muscle will increase.

Stress can vary both over a component (along a beam or over a plate or whatever) and through the component, from top to bottom, say. In the latter case, the variation is over a cross section, so at any beam cross section, the stress can vary from a point of maximum tensile stress to one of maximum compression stress, together with varying *shear* stress, and these stresses can all vary along the beam. Similarly, a plate can have stress variations through the thickness and over its extent. These types of stress will be explained.

But it is easier to explain first what *strain* is; the two are connected. Mechanical strain (often called *engineering* strain) is precisely defined as the change in a dimension of the structure due to applying a load divided by the same dimension when that load is not there; it is a ratio of the change in dimension to the original dimension. So for a beam in tension, as in the diagram below, it is the extra bit in dotted outline divided by the original length drawn with a solid line:

Strain = a deflection ÷ an original length

$$\varepsilon = {}^{d}\!/_{l}$$

The ε is the strain, d is the relevant deflection, and l the original dimension.

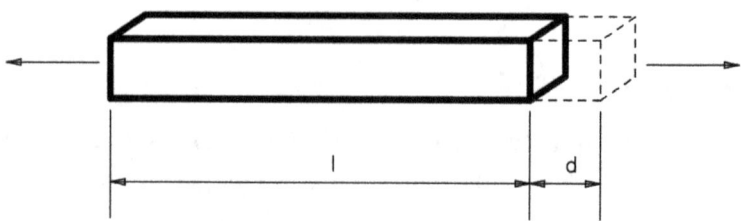

Figure 7.7 Extension of a simple bar under end load

These changes in a dimension are very small, usually less than one part in a thousand for metals, wood, reinforced plastics, and ceramics. The strain in rubbers and unreinforced plastics are, by contrast, perceptible to the human eye. These strains always occur. When a small feather alights on a sturdy oak table, the table sags under the weight of the feather, even if the table is made with a two-inch thick top and six-inch square legs; not by much, but sag it does. There is no such thing as rigid or immovable; every structure, however solidly built, always deflects under any load, however small it is.

If you purely stretch a beam, you put it into *tension,* it elongates; if you push on the ends of the beam, you put it into *compression,* it shortens. You can *bend* it. In this case, one side of the beam is in tension and the opposite side is in compression. The middle is unstrained.

Also, there is *shear* strain. You can picture this as follows: imagine cutting through a beam to produce two shorter ones; shear is sliding the two parts over each other on the cut faces. Obviously, real structures are not cut, but the action is the same; adjacent sections try to slide, in a parallel fashion, relative to each other. Bolts connecting two parts of a structure, when the load is at right angles to the bolt axis, are in shear. It is a concept that is a little abstract. In many structures, the three strains are connected, one often being caused by others.

Lastly, you can twist a body, producing *torsion.* The strain produced by this is principally shear. Adjacent sections of the body try to slide over each other in a circular manner.

As mentioned, there is a connection between strain and stress. Robert Hooke (1635–1703) first described, in 1678, the proportionality of the deflection of materials to the load causing it, Hooke's Law. Thomas Young (1773–1829) provided the actual figure of proportionality for steel, in a publication in 1807, by finding the frequency of vibration of a tuning fork. This he calculated as 29×10^6 lb per in^2 (200,000 MPa), which is still a good figure for modern steels. Again, it is called Young's Modulus. So stress is equal to strain multiplied by the Young's Modulus for the material in question. In equation form:

$$\sigma = \varepsilon \times E$$

It is usually manipulated slightly and quoted as Young's Modulus equals stress over strain, or as follows:

$$E = \frac{\sigma}{\varepsilon}$$

The letter σ signifies stress, ε signifies strain, and E is Young's Modulus. Note that each type of material has its own value of Young's Modulus which is intrinsic to the material and has initially been found experimentally. They vary in value by several orders. The values are published for a large variety of materials.

For each type of stress there is a formula involving the load and geometrical properties of the component. First we will consider the case of simple tension and compression stress, also known as *normal* stress or end load stress, then bending, shear, and torsion.

1. Pure Tension and compression

Bars and beams

Here the load is simply divided by the area of cross section. Each of the (imagined) tiny fibres carries the same load so the total load is uniformly carried by the total area. If we are considering a beam, say one of those steel posts used in a building – you see them everywhere a new large building is going up – then you must imagine that you have cut the beam at right angles to its length, and of the load, and calculate the area of the cut face. In this case, it will be the area of the three rectangles making up the cross section, as shown below.

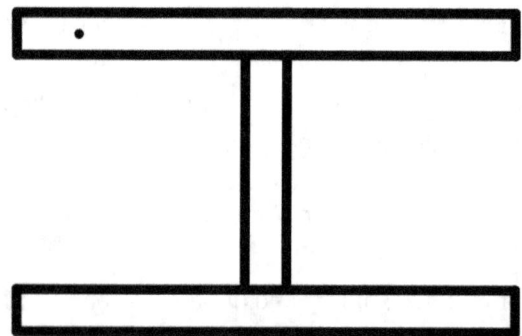

Figure 7.8 H-section beam made up of three rectangles

The area required is just the sum of the length times the breadth of each of the three rectangles.

The stress equals the load divided by the area:

$$\sigma = P \div A$$

The σ represents the stress, P the load, and A the total cross-sectional area. Conventionally, a positive result represents a tension stress, while a negative one represents a compression stress. Note that the load must be exactly at right angles to the plane of the cross section – that is, along the axis of the beam. If it is not, you must take that component of the load that is at right angles; the other component, parallel to the plane, will produce a bending moment and shear (see below).

The cross section can vary widely, from a simple circle, as in a bar; or a single rectangle, even in a width of plate; to a complex shape. This last can always be imagined to be made up of simpler shapes and their individual areas added together. In the case of a beam with a varying cross section, the stress obviously varies with this change, as in Figure 7.9 below. So moving from left to right, the stress has a certain value at the left end, staying constant up to the step, and then increases abruptly. It stays constant again until the bar starts to become deeper, where it reduces until the section becomes constant. But the local variation at the lug is ignored. This is not a strictly accurate thing to do, but the lug gives only a very small change in area, and the resulting stress variation is conservative.

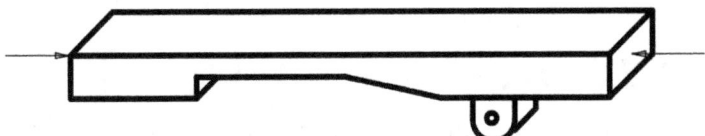

Figure 7.9 Variable section beam

The diagram shows the load putting the bar into compression; as mentioned previously, this would obviously shorten it. If the load, represented by the arrows, were reversed, the bar would be in tension and would lengthen.

Buckling

It has been stated that compression is just the opposite of tension, and certainly it is calculated in the same way, but this is only true up to a certain compression stress. At this point, the component buckles. The actual stress at which a component buckles, known as the *critical* stress, depends on its cross-sectional shape, material, size, and support.

The phenomenon is easily demonstrated with a flat foot-long plastic ruler. If you hold it between the palms of your hands and push them together, at a certain force level, the ruler starts to bend. This is buckling of a bar simply supported at each end. If you could hold the ends firmly enough to prevent them rotating, the ruler would buckle at a higher load. This shows that the end conditions are important. If the ruler is of steel, the critical force is higher. If the ruler is longer, the critical force will be lower. If the ruler is one of those three-sided ones and therefore much stiffer, the critical force will be much higher and probably the ruler will break without buckling.

When a component buckles, the stress situation is more complicated than pure compressive stress – in addition, now there is bending stress. The important question is whether the critical buckling stress is more or less than another allowable stress. If the buckling stress is more, then there is no need to worry about it and the stress calculation is indeed the same as in the tension case; if less it becomes the allowable stress. Buckling is discussed in Chapter 8 under allowable stresses.

Plates in tension

Plates can obviously be pulled into tension and are then just thin wide bars. They are treated similarly to bars. Plates are often loaded by pressure. If a flat plate has a pressure applied to its surface, it will initially react by bending, if the edge support conditions allow it, but when the stress reaches the *elastic limit* (see later), it will form a cylindrical surface so long as it does not fracture and then react to further pressure with tension. When a flat plate has a maximum deflection of about half its thickness, it starts to carry some load by this tension.

Vessels intended to carry a pressurized fluid are either cylindrical or spherical. This is an efficient way to contain fluids because the stress is far lower than in a flat-sided vessel, which would have to sustain bending stresses. In a cylinder, the largest stress in the skin of the vessel is the so-called *hoop stress*. This is directed around the circumference of the vessel at right angles to the axis. It is accompanied by a longitudinal stress at right angles to the hoop stress, directed parallel to the axis of the vessel. This stress is half the hoop stress. Of course, when you get to the end of the cylinder, things get more difficult. Most closures generate some bending. The hoop stress in a cylinder is

$$\sigma = \frac{pr}{t}$$

And the longitudinal stress is $\quad \sigma = \dfrac{pr}{2t}$

The stress in a sphere in any directions is $\quad \sigma = \dfrac{pr}{2t}$

This is the same as the cylinder's longitudinal stress. As usual, σ means stress, *p* pressure, *t* thickness.

Plates in compression

Here, as in bars, the stress starts as uniform through the thickness. However, especially if they are thin in comparison to their length and breadth, they easily buckle. The surface becomes quilted and the stress situation complicated.

2. Bending theory

In the case of bending, the stress is also calculated from a load and functions of the cross section. These are, however, more complicated than the case of simple tension. The load is now a *bending moment*. So first we have to find this bending moment (BM).

Beams

As mentioned in chapter 5 (e), BMs are the product of a force and a distance. These are calculated from one end or edge of a structure to the section in question. So starting from one end, for a beam, the BM at some point along its length is the sum of all the loads on it, including the reactions or support loads, times their individual distances to the point. This gives the tendency to curl up the beam which is resisted by the stresses developed in the beam. Obviously, the BM varies along beam as the distances vary. Some loads will be in one direction and others in the opposite, so the BM of one lot will reduce the BM of those in the other direction. The usual procedure is to draw a diagram of the varying moment along the full length of the beam. Once drawn it becomes obvious which section along the beam has the highest BM and ought to have its stresses calculated. This saves having to consider many sections. If the beam is of constant section only the position with the highest BM need be considered. If the beam varies in section then the worst combination of high BM and small section has to be sought out. Here the stresses will be highest.

A similar diagram for shear loads can be drawn and the two diagrams are usually constructed together. The figures below give some examples.

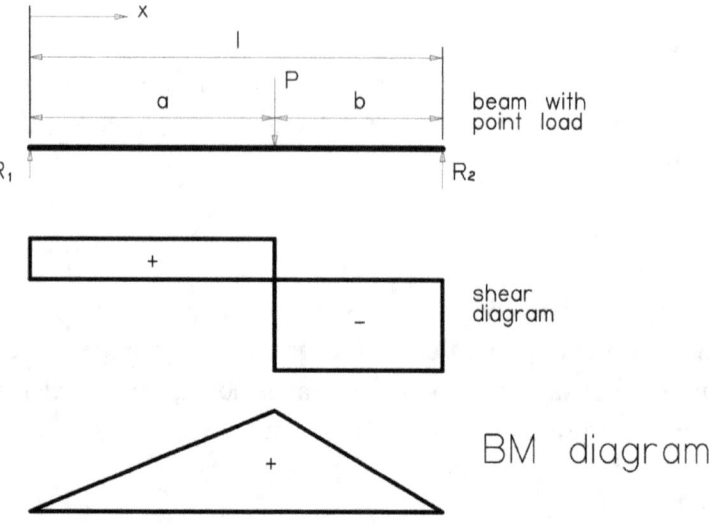

Figure 7.10 Simply supported beam with a non-central point load, its shear force diagram and BM diagram.

In order to construct the shear diagram you must first, as always, find the reactions, call them R_1 at the left end and R_2 at the right-hand end. If you take moments about the right-hand end, you find

$$R_1 (a+b) = Pb$$

Note that

$$a + b = l$$

then

$$R_1 = P \frac{b}{a+b} = P \frac{b}{l}$$

and from vertical balance $R_2 = P - R_1$

the meaning of the symbols are in the Figure.

Now, assuming that upwards loads or forces give a positive SF and downwards a negative one, you draw a horizontal line representing the value of the SF. It starts at a value of R_1 and only changes when another force is encountered, in this case P. P is larger than R_1 and in the opposite direction so the line moves down below zero to $-(P - R_1)$. It continues until the right-hand end, where R_2 moves it up to the zero line; simple really.

The BM is also drawn from the left hand end of the beam. The first force is the reaction R_1 and the distance from that to any point on the beam, say x, multiplied together is the BM.

So
$$M = R_1 \times x$$

When you come to P, its contribution to the BM has to be counted and now x is bigger than a so the BM becomes

$$M = R_1 \times x - P(x - a)$$

The minus sign before P is because P is directed downwards. Since the distances vary linearly and the forces are constant the BM also varies linearly, giving the result shown in the diagram. You could, if you wish, start from the right-hand end for constructing these diagrams. The picture is the same. Try it.

It is obvious that the maximum BM is at P. If the beam is of a constant cross-sectional shape, then this is the only place that needs be stressed. If there is a change to a smaller depth section, then although the BM is also smaller, the stress could be higher. Both places need stressing. It depends which combination is more severe.

The maximum BM for a beam with a single point load is when the load is exactly in the middle, i.e. $a = b$. Here the BM is as follows:

$$M = 0.25Pl$$

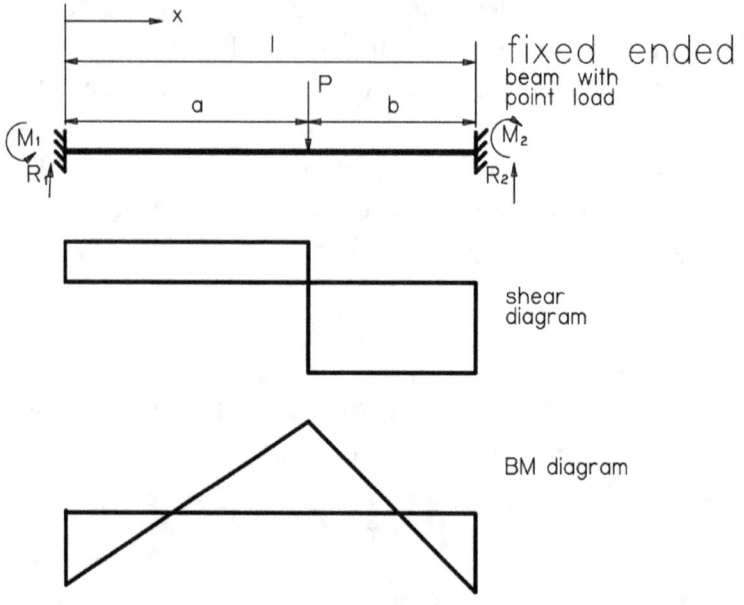

Figure 7.11 Fixed-ended beam with a non-central point load, its shear force diagram and BM diagram

First find the reactions. The structure is indeterminate, as you can see. If you remove any one of the constraints, the beam can still support the load, although the reactions will have different values. This requires equations involving deflections, which is outside our brief. The books, however, already give the answers.

$$R_1 = P\frac{b^2}{l^3}(3a + b)$$

$$R_2 = P - R_1$$

$$M_1 = P\frac{ab^2}{l^2}$$

$$M_2 = P\frac{a^2b}{l^2}$$

The directions of the loads and moments are as drawn in Figure 7.11. These are, intuitively, in the directions they would be. In complex loading, the chosen directions could be wrong, in which case their value would be negative.

There is now a moment at each end which alters the reactions, shear diagram, and BM. Because the reactions can be different in the fixed-ended case, the shear diagram will have different numerical values even though it looks much the same.

At any point x along the beam the moment is

LH end up to P $$M = -M_1 + R_1 x$$

P to RH end $$M = -M_1 + R_1 x - P(x - a)$$

It turns out that the maximum BM at the load, P, in the SS beam is greater than that in the fixed-ended one. At worst, with the load central, the moment in the centre is

$$M = 0.125Pl$$

And the highest end moment is when the load is at one-third of the length of the beam:

$$M = 0.1481Pl$$

These compare with the $0.25Pl$ in the SS beam. You can see that the BM diagram has, as it were, been pushed down so that now a moment appears at the ends and the moment in the middle is reduced; a lighter beam results. Take care, though. A true fixed end cannot necessarily be achieved in all circumstances – that is, all load cases. Often the true degree of fixity is between simply supported and fixed.

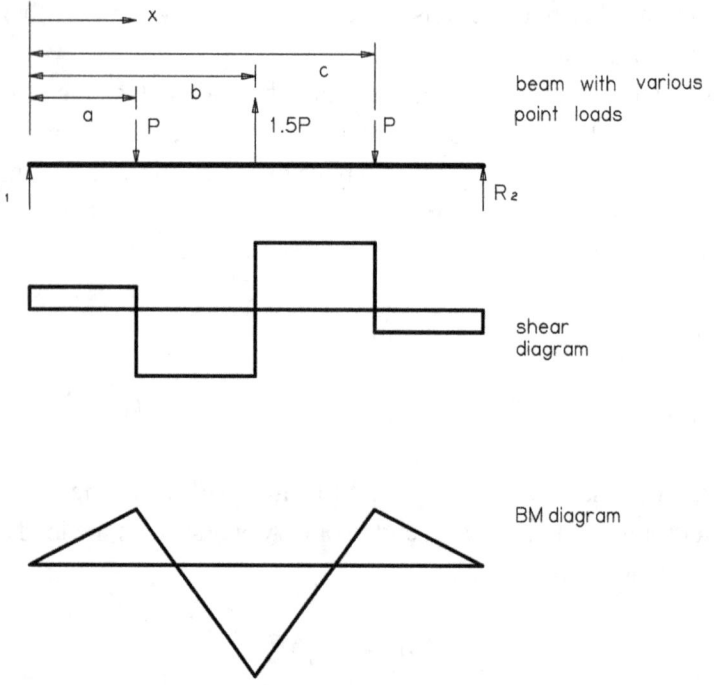

Figure 7.12 Simply supported beam with three non-central point loads, two down and one up, its shear force diagram and BM diagram

More complicated loadings are dealt with in the same manner. First the reactions; these, in the case in Figure 7.12, are the sum of the reactions for each individual load. Alternatively, just take moments about one end, as in Figure 7.10. This results in a more complicated equation and therefore one more prone to error. Then use the vertical balance equation to find the other reaction.

Next, the shear and BM diagrams can be constructed. The shear diagram is useful when the beam is in two parts and has a joint. The number of bolts required can quickly be calculated.

The beam in Figure 7.12 has more complicated loading which leads to more complicated shear and BM diagrams, but they are constructed in the same way. Then, working from the left end, the shear diagram is built up: R_1 up until the first load, which pushes the load down by the value of P; then to the second load, *1.5P* up; to the third P and down again; and finally R_2 up. The BM is, as before, the sum of the loads to the left of the position, x, times their distance from position x along the beam.

So the equations for the BM are as follows:

From R_1 to the left hand P $M = R_1 \times x$

From P to $1.5P$ $M = R_1 \times x - P(x - a)$

From $1.5P$ to the right hand P $M = R_1 \times x - P(x - a) + 1.5P(x - b)$

From right hand P to R_2 $M = R_1 x - P(x - a) + 1.5P(x - b) - P(x - c)$

If you are studying a similar situation, you can build up a similar set of equations and then substitute numbers for symbols and do the arithmetic. The worst BM will appear after this has been done.

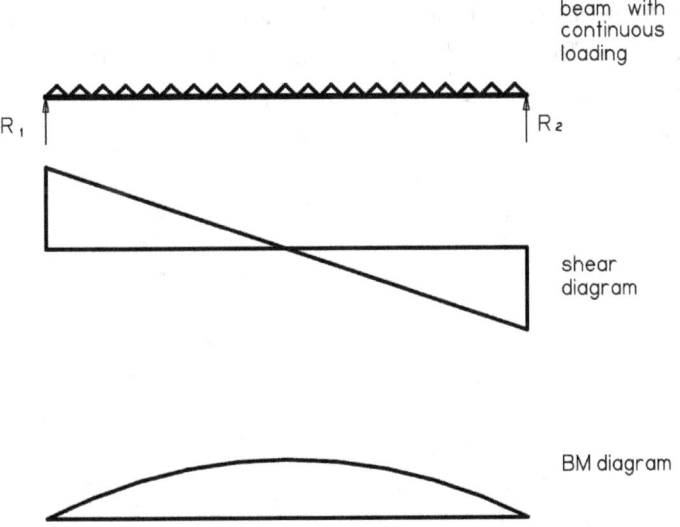

beam with continuous loading

R₁

R₂

shear diagram

BM diagram

Figure 7.13 Simply supported beam with continuous load, its shear force diagram and BM diagram

The wiggly line over the representation of the beam indicates a continuous loading on the beam. This gives a sloping shear diagram and a parabolic BM. Again, if the ends are fixed, the BM curve appears to be lifted up, moments appearing at the ends.

For a SS beam, the maximum BM is at the centre $M = 0.125wl^2$

For a fixed beam, the fixed moments at ends are $M = 0.08333wl^2$

Fixed-ended mid-bay moment is $M = 0.04167wl^2$

Many combinations of loadings and end conditions can apply to beams. Continuous loading can be over only part of the beam, and moments could be applied anywhere. Either end could be fixed, SS, or free. Each situation will have its BM diagram. As an exercise, try to construct the shear and BM diagrams for a cantilever loaded by a point load at the free end, then again with a continuous load.

When the structure is made up of several beams joined to each other at their ends, as in a building on columns, for instance, the moments at each support vary considerably depending on the load case and on the relative stiffness of adjacent beams. This type of structure is often known as a multi-span beam. The example below illustrates how the end fixity can vary.

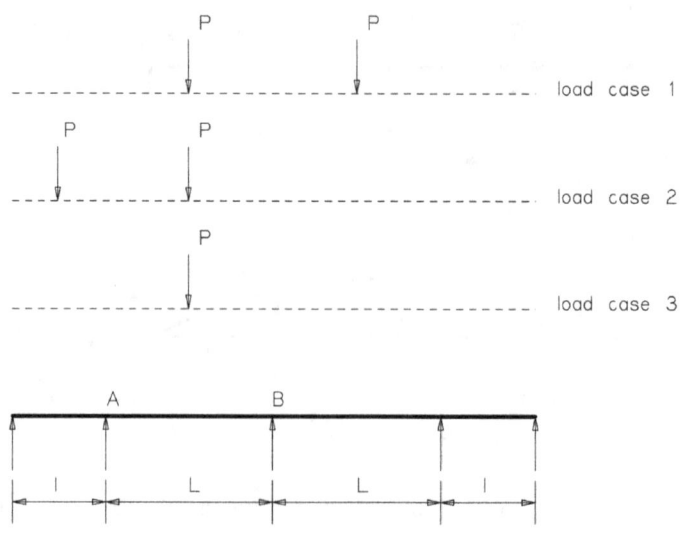

Figure 7.14 Four-span beam with three different load cases

This beam has four spans. The outer two are of equal length but are shorter than the centre two, which are also of equal length. We shall consider the degree of fixity at supports A and B. There are three different load cases which we shall examine one by one.

It looks as if the long bay, AB, is the one which is going to have the biggest bending moments because it will deflect more, which implies bigger moments. The three positions of interest are at the supports at A and B and mid-bay between A and B. Initially, we will assume that the entire beam is of constant section. This is usual when the structure is made from standard sections; the size of section would be decided by the worst stress anywhere along its length.

In the first case, we have two equal loads which are equally disposed about the central support B. The structure and its loading are therefore perfectly symmetrical about B. This symmetry means that you only have to consider half of it. If you were to do the analytical mathematics, you would find that the deflected beam was horizontal at B and curved down equally on either side. The slope to the deflected shape of the beam is zero at B. In turn, this means that the beam is fully fixed there.

A simple and acceptably accurate (though not precise) technique is to assume that AB is an isolated beam, fixed at B and SS at A, often known as a *propped cantilever.*

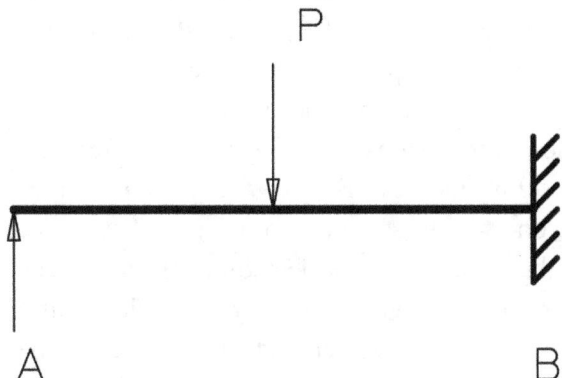

Figure 7.15 Propped cantilever extracted from the four-span beam
Figure 7.14

This ignores the fixity supplied by the outer beam, which reduces the BM in AB. You can then look up the BMs in a book; a few relevant ones are given at the end of this section. You will get somewhat higher values than necessary, but that is on the safe side. So the BM at B you find is $0.1875PL$. This compares with $0.125PL$ for a beam with both ends fixed. The true situation is evidently between these two figures. The BM in the middle of the beam, which is of opposite sense, is between the propped cantilever value of $0.15625PL$ and the fixed-ended beam value of $0.125PL$. These figures are for a mid-bay load. If you move one of the loads along a bit, then the BM at B is no longer exactly fixed; equally, if you increase the value of one of the loads, the BM at B changes.

In load case 2, the loads which are centrally disposed in their bays and are equal, but because the lengths of the two bays are unequal, the BM at A is not the fixed-ended one quoted in the books. The quick but conservative calculation is to add the BM at A for a propped cantilever for the short beam to that of the fixed-ended longer beam. If l is the length of the short span, L the length of the longer span, and the loads are in the middle of the spans, then the total moment is $0.1875Pl + 0.125PL$. For a quick estimate, the mid-bay BM in each case is the propped cantilever condition: $0.15625Pl$ or $0.15625PL$.

In the last case, the BMs at A and B are unequal because the lengths of the bays on either side of the loaded one are unequal and so provide unequal restraint to it. Here you will get a conservative idea if you assume the bay is a propped cantilever. If the load is central, the moments at B is $0.1875PL$ and mid-bay is $0.15625PL$.

In addition, if the beams in the bays are of different section (depth, shape, and so on), then the BMs at the supports will change again. Incidentally, one way to solve this problem to get an accurate answer is to equate the slopes to the deflected beam shape of each bay at the common supports. This will lead to some simultaneous equations, but you may not want to get involved and choose to go to an expert. There is another method called the Moment Distribution Method. This is a tabular technique attributed due to a professor called Hardy Cross. This is fairly complicated and not used much nowadays because computer methods have overtaken it.

The list below summarizes the results given for this example. They are not, however, the maximum BMs, for they assume the loads are exactly in the middle of their bays. The maximum BMs occur at different locations for propped cantilevers. Their results are also given.

Case	Position of moment			Notes
	A	mid	B	
1		0.15625PL	0.1875PL	
2	0.1875Pl + 0.125PL	0.15625PL + 0.15625Pl		
3		0.15625PL	0.1875PL	
Max possible		0.174PL		Load is 0.634L from fixed end
Max possible			0.1927PL	Load is 0.4227L from fixed end

Table showing moments in the three cases and the maximum possible for a propped cantilever

Now that we have found the worst BM in a beam under consideration, we can get back to the stress caused by this BM. The problem is that the stress produced by a BM varies across the section, as shown in the next diagram, which is of a rectangular bar cut at the section where the stress is to be calculated. Fibres at different heights carry different loads; therefore, the stress is different.

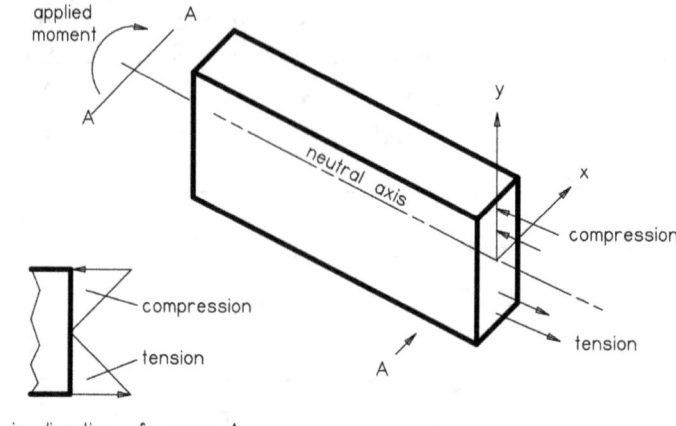

applied moment

A

neutral axis

y

x

compression

tension

A

compression

tension

A

view in direction of arrow A

Figure 7.16 Beam showing vertical stress distribution on a section due to a moment about a horizontal axis, AA, parallel to the x-axis

In this diagram, Figure 7.16, the maximum stress is at the extreme top and extreme bottom. Here the top-most or bottom-most fibres have the highest force on them. At the top, the stress is compression; at the bottom tension and in the middle, it is zero. The view in the direction of arrow A shows the full tension and compression fields, which spread fully across the section in the x-direction. The stress at any height is directly dependent on the value of the dimension y. It is said to be linearly distributed down the section. The formula for the bending stress is

$$\sigma = \frac{My}{I}$$

The letter σ is the stress, M the BM at the section, y the distance from the neutral axis, or NA (in this case, the centre), and I the second moment of area (see Chapter 7, section (a) (9), and the addendum).

There is a need to be careful about the relative orientation of the moment and the section. In Figure 7.16, if the moment were to act at right angles to the direction shown, then the highest stresses would be on the sides of the bar, not the top and bottom.

It must be realised that the bending moment, the orientation of the cross section, and the second moment of area must all correspond in the directional sense. See Figure 7.17.

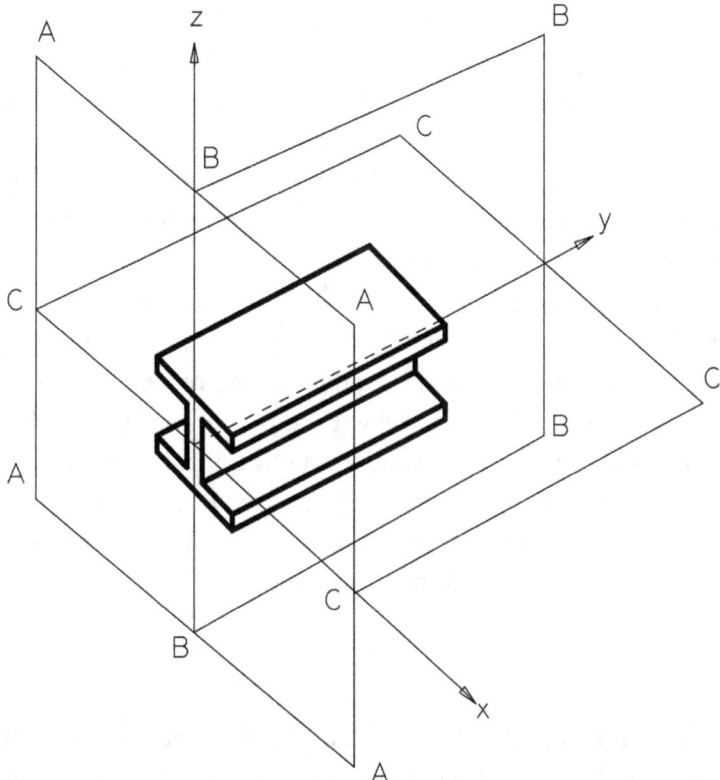

Figure 7.17 Three planes at right angles to each other, defining the orientation of an I-beam

Here we have an I-beam with three planes defining its orientation. The first plane, AAAA, contains the x-axis and z-axis; the second, BBBB, contains the y-axis and z-axis; and the third, CCCC, the x-axis and y-axis. Obviously, if the bending moment curls around just the x-axis, it will put the flanges of the I-beam into tension and compression. If it just curled around the z-axis, the left and right edges of the flanges would experience the highest tension and compression stress. A moment around the y-axis would twist the beam, producing torsion stresses.

If the applied moment does not lie purely in one of the planes, it is best to work out the components that do. Then calculate the stresses from each component and add them at each critical point. So if a moment, M, is about a line in the AAAA plane which is at thirty degrees to the z-axis, then M cos30 acts about the z-axis and M sin30 about the x-axis. Now you can look up the standard I values, which are always given in the books about these axes, and calculate the stresses and add them at each position of interest, taking account of their sign, whether they are tension or compression. The maximums will occur at two diagonally opposite corners of the beam where the two greatest tension stresses add and the two largest compression stresses add to each other.

It is important to notice the sense (tension or compression) of stresses in calculations. The allowable maximum values may be different and stresses from two sources may tend to add or to reduce each other.

The following table gives a few examples of the BMs at the ends and mid-bay of beams. There are many more comprehensively given by *Roark* (see bibliography). Remember, the formulae assume either zero fixity or exactly full fixity at the beam's end(s). The numbers, k, in the table are to be substituted in this equation:

$$M = kPL$$

M is the moment, L is the length of the beam between supports, P is the load; in the case of the point load, it is just that, but in the case of the distributed load, it equals the loading per unit length times the length of the beam.

$$P = wL$$

w is the loading per unit length (millimetres, inches, or whatever). This is known as the uniformly distributed load, or UDL.

supports	Point load at centre		UDL	
k				
	Centre UOS	end	Centre UOS	end
Cantilever	–	1.0	–	0.5
SS beam	0.25	0	0.125	0
Propped cantilever	0.15625	0.1875	0.0703 at 0.375L	0.125
Fixed-ended beam	0.125	0.125	0.0417	0.0833

Plates

Plates can also carry a BM, obviously. However, their cross section is simpler than that of many beams – the only geometrical variable is the thickness. The BM can vary not only along the length but also across the plate. So a rectangular plate, SS on all edges, both vertically and horizontally, under a pressure load has the highest stress in the centre. This drops to zero not only at each end but also at each side.

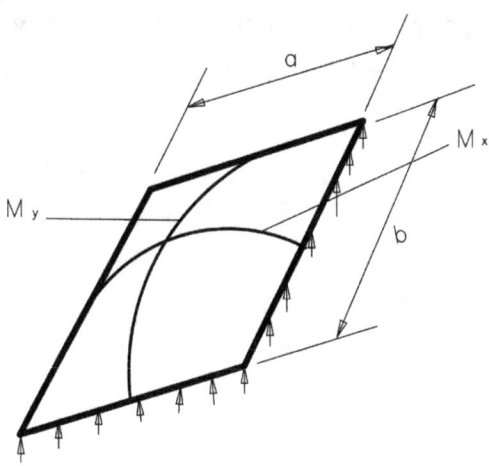

Figure 7.18 Rectangular plate with transverse loading such as pressure – SS all edges

The way that the BMs, M_x, and M_y vary across and along the plate is shown. You do not have to look up an I value, but you do have to look up a different factor depending on the length-to-breadth ratio of your plate if it is rectangular and remember, a square plate is a particular rectangular plate with length-to-breadth ratio equal to one. So the maximum stress for a rectangular plate is

$$\sigma = \beta \frac{pb^2}{t^2}$$

β is the factor you have to look up, p is the pressure, b is the breadth of the plate, and t is the plate thickness. There is a lot of mathematics behind β, and the books give the results. If you are interested in plates, Timoshenko's *Theory of Plates and Shells* presents many results for various shapes of plates under various support conditions and loads, as does Roark. Ignore the mathematics; look for the equation giving the stress in the relevant place. Remember, though, that the equations are all based on the idea of unit width. The results are given for points, and the adjacent points will have a different result.

   If the length-to-breadth ratio is three or more away from the ends of the plate, the plate acts as a beam across the short dimension of the plate. So a narrow strip of plate in this direction – again, say one unit (for example 1 mm) wide – can be treated like an individual beam; the stresses so calculated are a good approximation to the true ones.

Frameworks

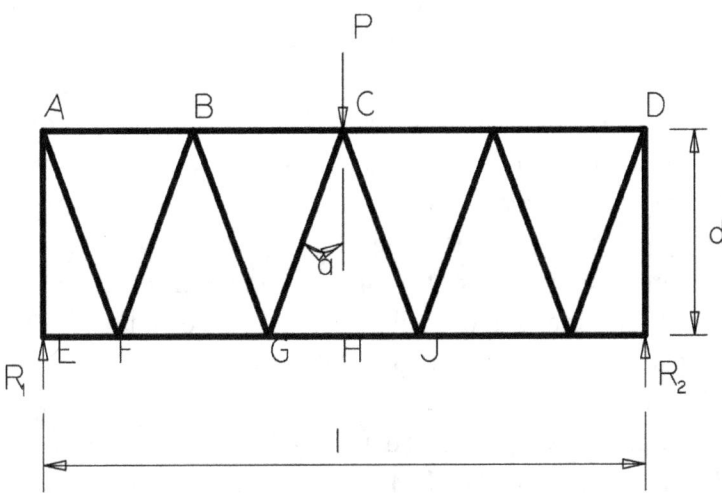

Figure 7.19 A simple framework with a point load at the centre.

Frameworks are an assembly of bars (called ties if they are in tension, struts if in compression) which can be used for roof trusses or bridges, for instance. If the loads are applied at the joints and the centre lines of the bars (strictly, the lines of the centres of area of the bars) meet at one point at the joints, then the bars can be considered to be pin joints since there are no BMs in them due to the end loads in them. Frameworks can be treated as beams in the first instance since together they form a rigid structure. We are considering structures such as *Figure 7.19*. We ignore the various individual bars and think of it as just a beam like the one in *Figure 7.10*. First, of course, we must find the reactions which is easy in this case as the load is symmetrically placed at the centre. Without further ado, we see that $R_1 = R_2 = P/2$. The BM diagram is as in *Figure 7.10* with the peak in the middle. The maximum moment is

$$M = \frac{P}{2} \times \frac{l}{2} = \frac{Pl}{4}$$

Next we notice that the geometry of the bars is symmetrical about the vertical centre line, which means that the load in a particular bar on the left of the structure is the same as that in its mirror image on the right. All we have to do is solve for the loads in the bars on one side and we have completed the job. If the loads were not symmetric, this would not be true. The next thing we notice is that the structure is determinate (see Chapter 5, section (c)); that is, if any bar in the structure we are considering is removed, the structure collapses, providing that each joint is taken to be a hinge. This is also called a simple structure. Frameworks are obviously an economical way of constructing large structures.

The question is, how do we deal with the BM? How do we find the loads in each bar due to it so that we can specify the bar's size? It is fairly obvious that the top will be in compression and the bottom in tension, just like the beam in Figure 7.10. The actual load in the bottom bar, GH, can be derived by considering a vertical section at joint H, which is midway between G and J, directly under C. Here we divide the BM by the depth, d. The BM is now represented by two equal but oppositely directed horizontal loads, one at C and one at H. Horizontal balance is satisfied.

This divides the frame into two halves, and we will solve the loads in each bar in the left half. The right half is represented by the horizontal loads describing the BM and an upwards reaction to the applied load P at R_1.

This technique, of representing a part of a structure by the loads it applies to the rest is used in many circumstances.

The load in bar GH is
$$\frac{M}{d} = \frac{Pl}{4d}$$

The load in CG is $P/(2\cos a)$ where a is the angle between HC and CG at C, as shown.

All the diagonal bars have the same load in them, as can be seen by considering vertical balance at each joint. CG is in compression, GB in tension, BF in compression, FA in tension again. AE is in compression and equals R_1.

We can now solve all the loads in the horizontal bars by the same method – work out the BM at each joint and divide by the depth – and that is the load in the bar opposite the joint in question. Try it. You should end up with the picture below.

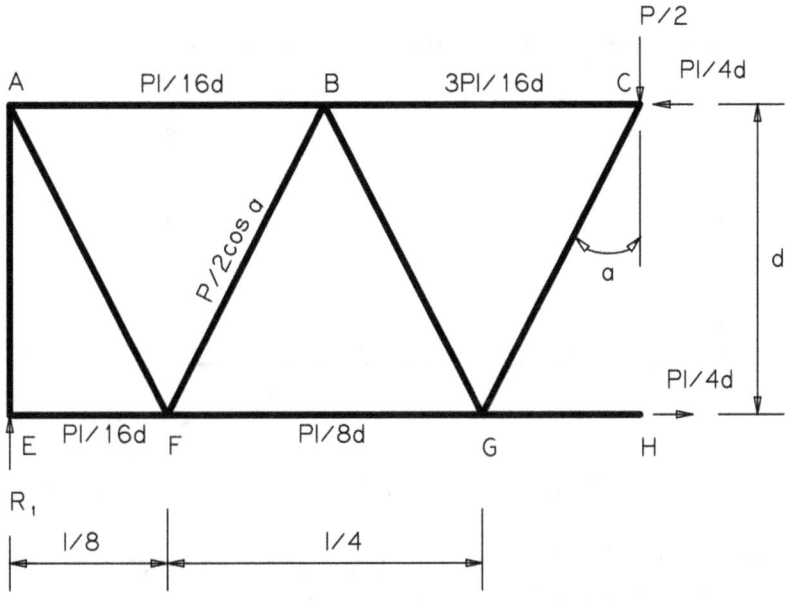

Figure 7.20 The left half of the framework with its loads

3. Shear

You can picture this as follows: if you imagine cutting through a beam to produce two shorter ones, then the action of shear is sliding the two parts over each other on the cut faces. Obviously, real structures are not cut, but the action is the same; adjacent sections try to slide, in a parallel fashion, relative to each other. Bolts connecting two parts of a structure, when the load is at right angles to the bolt axis, are in shear. It is a concept that is a little abstract.

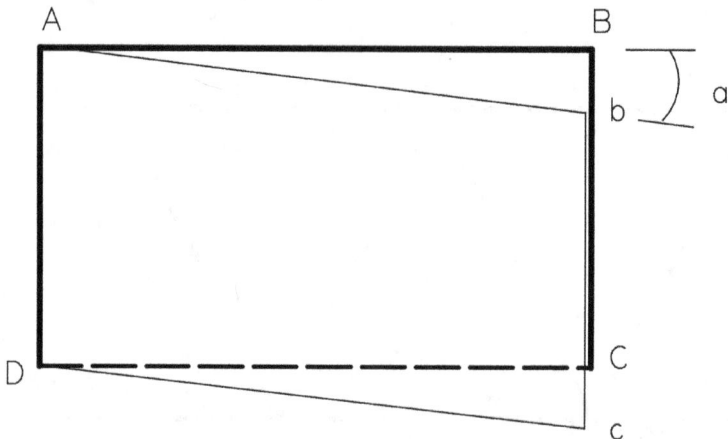

Figure 7.21. Panel in shear from ABCD to AbcD; shear angle a

The resulting deflection of a rectangular panel in shear is shown in Figure 7.21. The rectangle becomes a parallelogram in the same plane as the original rectangle, and point B moves to point b, and C to c. The angle, a, is a measure of the shear strain.

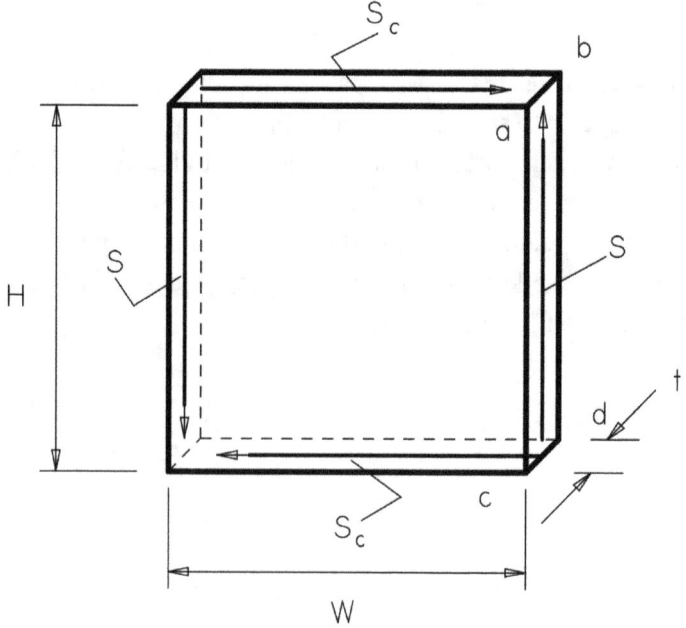

Figure 7.22 A shear panel with loads applie;. height H, width W.

Figure 7.22 shows another panel with shear forces applied and two edges showing, to make it easier to demonstrate how to calculate the shear stress. This is the force S divided by the area abcd. Call this A_s and τ the shear stress:

$$\tau = \frac{S}{A_s}$$

The shear area is

$$A_s = H \times t$$

Another point to notice is that the application of a shear load, S in this case, necessarily raises another shear load, S_c, at right angles to S. This is known as the *complementary shear*. The balance equations apply, obviously:

$$S \times W = S_c \times H$$

Bending also induces shear in a beam. The compression on one side of a beam in bending is balanced by the tension in the other side, and the transfer of the loads is accomplished by a shear stress in the material between. An I-section beam such as Figure 7.17 is arranged so that the tension and compression are mostly carried by the flanges and the shear by the web connecting them. Clearly, the web also has some tensile or compressive stress in it; the stress state is quite complicated.

There are a few situations where a section through a structural member is in a state of *pure shear*, but most of the time structures have a mixture of shear and tensile/compressive stresses. However, often one dominates and the other may be ignored. This is true of beams, which should be about ten times longer than they are deep for this to be so. The bending stress in them is nearer to its allowable limit than the shear stress is to its limit, by a considerable margin. Little error is made by ignoring the shear stress.

But if the beam is short in comparison to its depth – the length-to-depth ratio is only two or three, say – then the shear strength capability of the beam may be as big as the bending strength. The deflection of the beam will be lower than expected from a calculation which assumes all the load is carried in bending since some is carried in shear. The distribution of load between bending and shear is estimated by comparing the bending and shear stiffnesses. This type of beam is sometimes known as a *short beam*. The beams over the columns of ancient Greek and Roman temples are short beams, as stone has little tensile strength, therefore little bending resistance, and the loads on them are mainly resisted by shear.

Relatively thin metal shear panels can buckle. The buckles are at an angle of about forty-five degrees to the shear load. This may not mean that the structure has failed, as the metal's stress may be elastic – that is, they disappear when the load is removed. This would probably only be allowed deliberately in high-tech structures such as aircraft, when it results in lower weight.

4. Torsion

Another way for a structure to resist a load is by an action called torsion. This is a twisting of the structure along its length; for instance, applying a screwdriver to a wood screw. There need be no push or pull on the screw, nor any bending action. Adjacent sections of the screw try to rotate relative to each other, causing shear stresses. Of course, the structure can be anything from the simple example of the wood screw, or the screwdriver, to a complex one such as an electricity pylon.

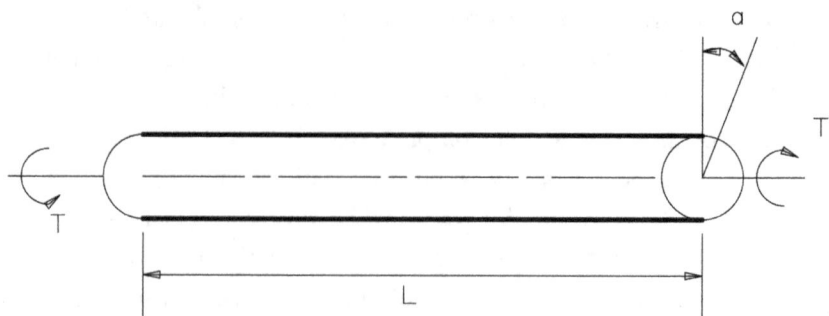

Figure 7.23 A bar under pure torsion

Figure 7.23 shows a bar under torsion. This is the simplest to analyse for the stresses and the angle of twist, which is developed in reaction to the torque. The equations for the angle of twist and the stress are as follows:

$$angle\ a = \frac{32\ TL}{\pi d^4 G}$$

Therein, a is the angle in radians, T is the torque in Nm, L is the bar's length, d is the diameter of the bar (both in m), and G is similar to E in the Young's modulus. Note that there are 2π radians to 360 degrees and that G is the ratio between the *shear* strain and the *shear* stress; this is in contrast to E, which is the ratio for tensile and compressive strain and stress. The two are connected.

The shear stress, τ, is given by

$$\tau = \frac{16T}{\pi d^3}$$

This stress is a maximum at the surface of the bar, reducing to zero at the centre. This implies that the middle part of the bar does not do much work, so it is more efficient to use a cylinder. The required equations for a cylinder:

$$angle\ a = \frac{32TL}{\pi G(d_1^4 - d_0^4)}$$

Therein, d_1 is the outer diameter and d_0 the inner diameter of the tube.

$$shear\ stress, \tau = \frac{16T d_1}{\pi(d_1^4 - d_0^4)}$$

Other cross-sectional shapes of bars and tubes need more complicated analyses. As usual, results for these can be found in reference books such as Roark. Non-circular shapes have a tendency to *warp*. This is a distortion of the cross section perpendicular to its face. So in a square tube, two diagonally opposite corners will lengthen and the other two will shorten – unless of course the end faces are restrained. In this case, extra stresses will be generated. Warping can be easily demonstrated by twisting a rolled-up piece of card. You find that the free corners of the card slide longitudinally relative to each other.

Open sections such as angle or channel sections are much less resistant to torsion. This is obvious if you handle such sections which are a few centimetres wide and compare them with a 1.5 cm copper or even plastic plumbing pipe. Sometimes a different reaction system resists the torsion. For instance, in an I-beam, the top and bottom flanges can develop differential bending so that one flange bends in its plane in one direction and the other in the opposite direction.

Note that structures can buckle under pure torsion. You can do it to a drinks can by hand. Long drive shafts can suffer a phenomenon called whirling. This is actually a bending mode, though the load is torsional.

5. Addition of stresses

It often happens that several load cases occur together. Each produces stresses in the structure, and these must be added to produce the total stress. Also, it has to be realised that stress has both magnitude and direction. When two stresses are added together, they must be in the same direction. By this, we do not just mean tension and compression, where, of course, the stresses tend to cancel each other somewhat. It is also important that they are not at an angle to each other. If they are, then only the parallel components add together. However, this leaves the component at right angles to consider; what does one do with that? In this case, the stresses are said to be *orthogonal* and special procedures apply.

We start simply, by considering one-dimensional cases; all the stresses are in the same direction. This applies to bending stress in beams, for instance, where all the major stresses are along its length. Simple tension stresses in plates, all in the same direction, also constitute a one-dimensional situation. Indeed, this applies to any solid body which has all its stresses in the same direction. The stresses due to all the simultaneous load cases can simply be added algebraically, taking notice of the signs of the stress.

Two-dimensional cases often occur in plates where several loads are applied in different directions. There may also be a shear load applied in the plane of the plate. In this case, we have to find what is known as the *principal stress*. Some examples in two (and three) dimensions are given in Roark's book. These all involve tensile (or compressive) and shear stresses in increasing complexity. They have to all be mutually at right angles to one another, so you may first have to resolve your stresses to achieve this.

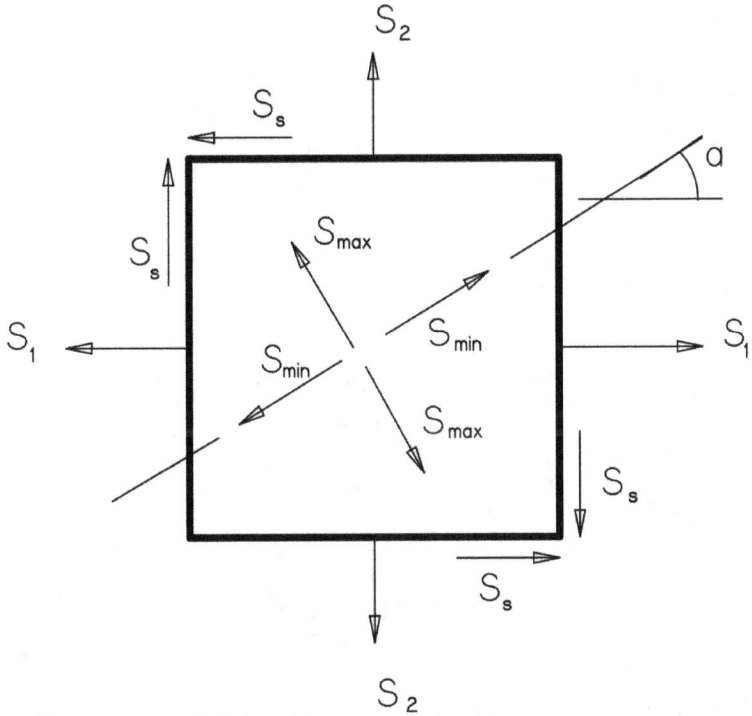

Figure 7.24 Two-dimensional stress system – biaxial stresses plus shear.

Imagine an infinitesimally small square of plate in some structure and you have calculated the stresses on it. This is equivalent to finding the stresses at a point. You have S_1 in the x-direction and S_2 at right angles to it in the y-direction. There is also an applied shear stress, S_s. Because these stresses interact, the maximum stress in the plate is not necessarily S_1 or S_2. The actual maximum is the principal stress which comes with a minimum stress; these are S_{max} and S_{min}, there being no shear stress at this orientation

The angle, a, defines the direction between the calculated stresses and the principal stress. This is the most complex two-dimensional case, and the equation for the principal stress is given below:

Tensile/compressive stress is

$$S_{max} \text{ or } S_{min} = \frac{1}{2}(S_1 + S_2) \pm \sqrt{\left\{\left(\frac{S_1}{2} - \frac{S_2}{2}\right)^2 + S_s^2\right\}}$$

This is at angle a:

$$a = \frac{1}{2}\arctan\frac{2S_s}{S_2 - S_1}$$

The maximum shear stress is

$$shear\ stress = \sqrt{\left\{\left(\frac{S_1}{2} - \frac{S_2}{2}\right)^2 + S_s^2\right\}}$$

This is at angle:

$$a = \frac{1}{2}\arctan\frac{S_1 - S_2}{2S_s}$$

The angle is at forty-five degrees to the plane of the principal stresses.

This is getting complicated, and it's still in only two dimensions. You will see that because of the plus/minus in the first equation, there are two values of principal stress. Accompanying the maximum principal stress is a minimum one. It may be important, especially if it is compression and buckling is an issue.

The first equation can be examined and used for simpler applied loadings. If the applied shear stress is zero, the principal tensile stress is equal to the highest of the stresses. Even though no shear stress is applied, one will appear at an angle to the line of the tensile stress. In the case of just one tensile stress, this is a maximum of half the tensile stress and is at forty-five degrees to the line of the principal stress.

In the case of one tensile stress and another equal in size compressive stress, at right angles to the first, the shear stress equals the tensile stress. Since allowable shear stresses are usually much less than allowable tensile stresses, this is a severe case.

In three-dimensional cases with several stresses applied, the principal stress may not be on the surface of the component; it may point into the interior. This situation can become very complex and is probably best left to experts.

There are other combinations of the stresses in a component to do with predicting its actual failure under some set of loads. Von Mises stresses are sometimes quoted as a failure stress in Finite Element programs (see later). They have to do with the energy of distortion. This is outside these discussions, which are concerned with a reasonably accurate idea of stresses that are not the most accurate possible.

These principal stresses are the ones to be compared with the allowable stress pertaining to the material.

6. Joints

It is very important to assess the strength of joints, as they are often points of failure in a structure. If not designed carefully, they become the weak points. There are a number of types of joint: bolted, riveted, welded (in metals and plastics), soldered, bonded (glued), merely by trapping two components together by some sort of interference between them and, most horribly, by friction.

Nuts and Bolts

There are numerous types of bolt, not just different sizes but also head types, thread types, materials, and so forth, all strictly specified and controlled by various British Standards agencies.

Bolts can have nuts or be turned into prepared or unprepared holes; these are often termed screws, if you want to be technical. The thread can reach right up to the head or there can be a length of plain shank. The type of bolt used depends on the integrity required and the type of load to be transmitted.

Bolts have a basic problem, and that is the fact that the threads are a helix, as they have to be, and so are not at right angles to the axis of the bolt. This means that there is a component of the tensile load, generated by the slope of the helix, which has an undoing tendency. If a nut on a well-oiled and smooth thread is shaken up and down, it will undo under its own weight. The answer to the problem is to tighten the nut hard. The friction between the threads and between nut and component has to be enough to counter the induced slope load. In addition, it may be decided to *lock* the nut (see later).

Bolts in shear and bearing

For bolts in shear, the highest integrity, giving the least likelihood of failure, is produced by using bolts with a plain shank going through the components. The holes in the components have to be drilled and *reamed* together on final assembly. Reamers have cutting flutes along their length, and they ensure that the bolt holes are exactly lined up and the precise diameter needed for the bolt to fit tightly. This is a more expensive procedure than the general engineering one of just drilling holes of various fits.

If, as usual, the bolts fit less than exactly in the holes of the components, then the load may be transferred between them by friction under the bolt and nut heads. This means they must be done up tightly. This system is used in steel frame buildings, which have especially stiff washers to increase the friction area under the bolt. If the assembly is subjected to vibration or shock loading, this friction may break; then the bolts and holes will contact each other and transfer the load that way by shear.

There are three modes of failure of shear joints. Firstly, obviously, the shank of the bolt may fail because the shear stress is too high. Secondly, the bolt hole may fail in *bearing*. Here the hole becomes elongated permanently – that is, the metal deflects beyond its elastic limit. It is possible that the bolt shank, if it is of a softer material, suffers this instead of the hole. The third type of failure is bending of the bolt. This can happen if the bolt is long compared to its diameter, often if there is a gap between the components as in a *lug*. The diagram shows some typical shear joints.

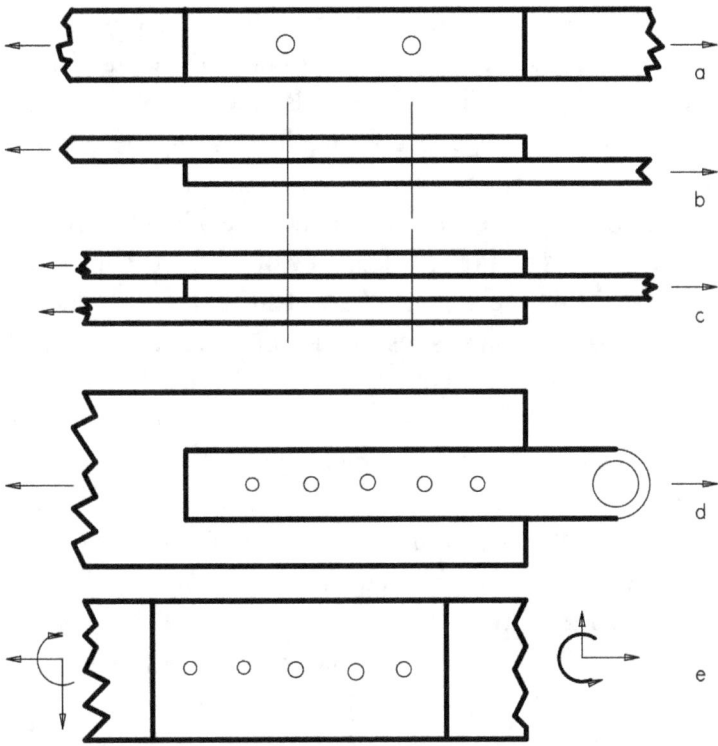

Figure 7.25 a) Simple two-bolt joint; b) in single shear; c) double shear; d) multiple bolts; e) a joint in a beam with multiple loads.

Figure 7.25a above shows the simplest type of joint, two components and two bolts in single shear. Single shear, as in Figure 7.25b, obviously, means there is only one bolt cross section where the load is transferred, while double shear, as in Figure 7.25c, has two sections for the load so the bolt shear stress halves and the load divides equally between the top and bottom straps. Incidentally, it is poor practice to have a single bolt in such a joint. Any vibration or sudden loading will tend to undo the nut, and then the joint fails. Two bolts double the chances of rescuing the situation before complete failure. Furthermore, if there is any transverse component to the load, the components will try to rotate about the bolt, and this will have a tendency to undo the nut if the load is in that direction. So always try to have two or more bolts in a joint.

The single shear joint has another problem. If you look at 7.25b above, you will see that the loads at each end are not in line. This means that the components will rotate and bend in such a way as to reduce the offset. The section at the bolt centre line is the least stiff and will therefore bend the most. Since there is also a *stress concentration* (see later) at this point, the danger of cracking in the component is high here, especially if there is a changing or *alternating* load. Fatigue failure may occur. This situation does not occur in the case of double shear.

In the two bolt situation, the load obviously divides into two equal bolt loads, provided the components are of equal thickness and the bolts are the same size. The shear stress in the bolt is the load divided by the cross-sectional area of the bolt. The area to be used is the core area (the area measured from the bottom of the thread groove) if the bolt of the components is threaded at the mating face. If not, the nominal area of the bolt is used. This can be crucial, as some threads are cut deep into the bolt diameter. The minimum diameter for the various types of thread can be found in British Standards, *Machinery's Handbook*, or manufacturer's literature.

In figure 7.25d, there are five bolts. The conventional, simple calculation is to assume that each bolt carries one-fifth of the load; however, the load in each bolt is not equal. The bolts at each end are the most heavily loaded. This is because of the flexibilities in the structure: the components, which may not be identical in size or material, stretch between bolt holes, the bolts deform slightly, and the holes give a little in *bearing,* all this resulting in the end bolts taking more of the load. As a rule of thumb, you can multiply the average load by one and a half. This means that increasing the number of bolts in a joint does not necessarily reduce the individual bolt load proportionately. If there are more than about six bolts, the end bolts may take more than one and a half times the average load. It may be time to seek expert advice if your joint is of questionable strength.

There is an assumption here that all the bolts fit perfectly in their holes. Obviously, if one has a hole that is too large, it will not bear and not take up its load, although for lightly loaded holes whose bolts are tightly screwed up, the load will be transferred from component to bolt by the friction under the bolt head and/or nut.

Figure 7.25e shows a bolt group to which a tension load and a transverse load as well as a moment have been applied. In this type of situation, which can get quite complicated, the first thing to do is find the *centroid* of the bolt group. In this instance, it is obviously the centre bolt, provided all the bolts are the same size. The horizontal load is initially divided equally among the five bolts; these act horizontally. Similarly with the vertical shear load. Since for each of these bolts the loads are at right angles to each other, the resultant is given as follows:

$$Resultant = \sqrt{v^2 + h^2}$$

$$angle\ between\ horizontal\ and\ resultant = arcsin\frac{v}{h}$$

Therein, v is the vertical load on a bolt and h is the horizontal load on that bolt. This is how to find such resultants and is, of course, due to Pythagoras.

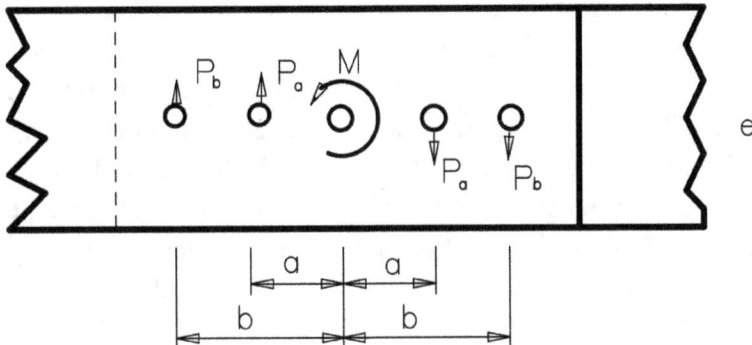

Figure 7.25 part e Bolt loads for a moment applied at a joint

The moment can be assumed to rotate the joint about the central bolt as above. The basic assumption is that a bolt is loaded in proportion to its distance from the centroid, so the ratio of the bolt loads is the same as the ratio of their distances from the centroid. The farther away from the centroid the bolt is, the bigger the load's value. The applied moment is reacted by the sum of the bolt loads times their distance from the centroid.

$$M = 2P_a \times a + 2P_b \times b$$

The ratio is

$$\frac{P_a}{P_b} = \frac{a}{b}$$

Manipulate these two little equations and you get this:

$$P_a = \frac{a}{a^2 + b^2} \times \frac{M}{2}$$

$$P_b = \frac{b}{a^2 + b^2} \times \frac{M}{2}$$

The 2 in the formulae is because there are two P_a's and 2 P_b's. In general, this is the format for any number of bolts. See the next example.

If the placing of the bolts is irregular, the centroid's position may not be obvious and will have to be estimated. For a quick answer, it may be possible to guess an approximate position, perhaps by ignoring a small bolt or just one irregular bolt. Then, as a test of the guess, draw a horizontal line and a vertical one through the centroid on a scale drawing of the arrangement (you may more easily do the trigonometry). Then the sum of the distances of each of the bolts above the horizontal line should equal the sum of the distances of each of the bolts below the line. Also, the sums to the left and right of the vertical line should be equal.

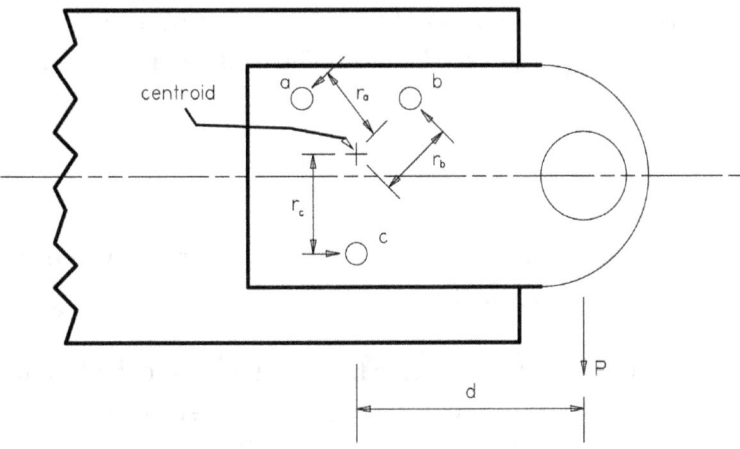

Figure 7.26 Lug held by three bolts, a, b, c, with vertical load P

In Figure 7.26, the centroid of the three bolts has been established and the distance of each bolt from it calculated as r_a, r_b and r_c. The direct vertical load is $P/3$ on each bolt, but there is also a load on each due to the moment Pd. Their reactions to the moment are directed in a tangential direction about the centroid as shown.

$$M = P \times d$$

The value of each bolt load is given by

$$\textit{bolt load a, } q_a = \frac{Mr_a}{r_a^2 + r_b^2 + r_c^2}$$

This is similar for bolts b and c. This assumes that all the bolts are of the same size. If they are not, each bolt cross-sectional area must weight each r. The equation above becomes as follows:

$$q_a = \frac{MA_ar_a}{A_ar_a^2 + A_br_b^2 + A_cr_c^2}$$

The A's are the cross-sectional areas of the relevant bolt.

Now the loads have to be added. For instance, there is a vertical load on bolt c of P/3 and a horizontal one of q_c. The resultant is

$$\sqrt{(P/3)^2 + q_c^2}$$

Do the same for the other bolts, except that the horizontal and vertical components must be found first.

When two plates are connected together by a row of bolts, the holes should not be too near the edge of the plates. A distance from the edge of the plate to the hole centre of one and a half times the hole diameter should be maintained as a minimum. The distance between hole centres should be at least twice the hole diameter. If these rules of thumb are broken, careful consideration by an expert should be sought.

The bearing stress on the surface of the hole in the component and on the matching surface of the bolt is given by

$$bearing\ stress = \frac{P}{dt}$$

P is the load, d the bolt and hole diameter, and t the thickness of the component.

Bolts in tension

Bolts in tension should be tightened when a structure is assembled to avoid fretting between the mating surfaces when loads are applied. Friction between the surfaces, however, does not play any part in transferring the load. The stress in the bolt is due not only to the applied load but also to this initial tightening. This is explained in Chapter 6 section (g), and Figure 6.4 is reproduced below to remind you of the loads present in the joint.

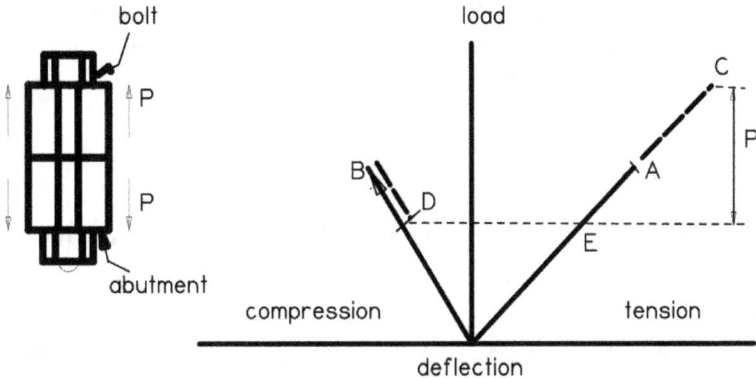

Figure 7.27. Deflections and loads in a pre-tensioned bolt

When calculating the strength of the bolt, the minimum cross-sectional area should be used, and this is usually between the bottom of the thread grooves. This is known as the minor diameter.

Sometimes extra thick washers are used to spread the bolt load over a larger area of component, especially if these are of a soft material. This avoids impressing the nut and bolt head shapes into the surface. It also increases the circumferential shear area of the components.

Reinforced plastics can be bolted in similar ways to metals. When bolt holes are drilled in components, there is a tendency for the layers of reinforcements to delaminate when stressed. This must be countered by screwing up the bolt sufficiently tightly and using washers to contain the delaminating force.

Locking of nuts

As mentioned, nuts inherently tend to come undone and so should be locked in some way if the joint is at all important. There are many ways of doing this, some more effective than others. Better methods include the following: putting a split pin through a castellated nut; two nuts tightened onto each other; using a bonding chemical on the threads; using a nut incorporating an insert, such as nylon, with a hole which is of smaller diameter than the bolt. It is also possible to peen protruding threads of the bolt, which ruins them, so this is a permanent method. Special so-called shake-proof washers are available.

In some circumstances, using a very high tightening torque is adequate. Experience in particular industries and circumstances are often the basis of their practice.

Rivets

Rivets do not have nuts to close one end but have the end deformed, with a specially formed anvil and hammer, so that they will not pull out of their holes. They are thus permanent and cannot be undone without destroying them. Rivets, which should only ever be used in shear, are of two basic types: protruding head or countersunk head. The strength of protruding head types can be calculated in the usual manner. But because the failure mode is complex, involving gross distortion of the head, countersunk rivet strength is established by test. The strength figures are available in the literature and are, very approximately 70 per cent of the protruding head value.

The strength varies with the thickness of the plates, and if the plates are relatively thin, then the critical strength may be bearing in the plate hole. This must be checked.

The edge distance, as with bolts, must be maintained.

Blind fasteners

There are large varieties of bolts and rivets which can be set from one side of the joint only. This means that components with holes which are unreachable from one side can be fixed together. Often a special tool is required to complete the joint, which may involve deforming a collar on the blind side and breaking a central pin. This effectively, from the strength aspect, leaves a cylindrical rather than a solid shaft so that the tensile and shear areas are reduced. The actual strengths are found by testing and are available in manufacturer's literature.

Welds

At first sight, welding two pieces of metal together so that you cannot see the join seems to be the best thing since sliced bread. However, there are problems. The metal at the join has to be melted and the two pieces pushed or held together. Apart from the difficulties involved in heating the pieces locally without destroying them, there are two major effects. First, the metal's microscopic make-up (that is, its grain structure) and its composition are changed, and this means its strength properties are changed, usually adversely. The tensile strength can be largely recovered by heat treatment, which means reheating the component to a temperature much less than melting but still so hot that special facilities are needed. Some steels have been developed that do not lose much strength for special applications, and some metals or some of their alloys cannot be welded at all.

For quick estimates, assume that the strength is reduced to 85 per cent of the published strength of the steel. If this causes difficulties, experts will have to be consulted.

The second major point to be considered is that welding often distorts the components. This is because the part of the component that is to be welded is heated. The resulting temperature gradients cause local stresses that may exceed the yield strength of the material, resulting in permanent distortion. If important, these distortions will have to be corrected. Sometimes the local stresses are ignored – they are, after all, local – and this may be acceptable if there is no fatigue loading. If there is and the cyclic load adds to the welding, induced-load failure will occur after a time. This, however, is a difficult situation to analyse and needs the attention of specialists.

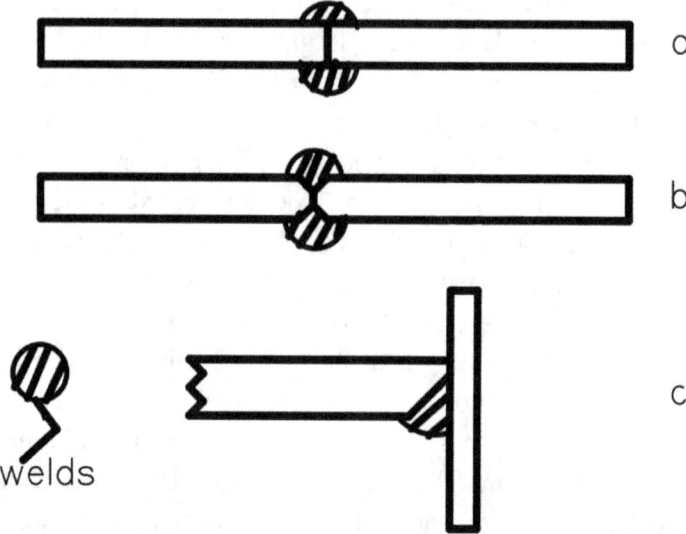

Figure 7.28 Cross sections of some welds

Apart from the reduced strength of the weld material, there is the question of available cross-sectional area. For instance, a simple end-to-end butt joint between two components must have sufficient weld area so that its strength is enough. The area of raw weld is difficult to assess, so they are often ground smooth, but then there may not be enough area. This is shown in Figure 7.28a, where too much grinding will reduce the strength of the joint. The weld edges should be prepared by being chamfered, as in Figure 7.28b; this ensures that the weld area is not completely ground away. Similarly, in Figure 7.28c, a nice radius can be ground into place – good for fatigue avoidance. In the case that the weld is not ground, care must be taken that the material assumed in the calculation is present – welds are not necessarily uniform all along their length. Care must be taken in assessing their dimensions.

Most welding uses a filler rod to add material to the joint. This may not have the same strength as the parent material. Some processes do not add material – for instance, spot welding. Here the weld takes place under pressure; the problem is that a permanent stress exists in the weld when the pressure is removed. This may be important.

There are similar jointing methods to welding: brazing and soldering. These are lower temperature methods which do not melt the parent material. Brazing is done at temperatures above about four hundred degrees Celsius, depending on the filler material. Usually the components are pressed together and the brazing material fills the gap by capillary action. Soldering is not usually considered to have any strength. It is used to seal joints or provide electrical contact. In the electronics industry, small components are supported by their welds, but this is a special case.

Unreinforced plastics can also be welded. There are a large number of methods, mostly requiring special tools. As in metal welding, the components are melted at the joint and pressed together. The heating is by hot air guns, ultrasonics, lasers, friction, and so forth.

The strength of a plastic weld is dependent on the method. If the method allows the long chain polymer molecules to intermingle sufficiently, then the basic strength of the parent material ought to be recoverable. Shortfalls can be compensated for by increasing the contact welded area. Test results should be used to check that the desired strength has been achieved.

Bonding

Most materials can be bonded together, and the components need not be of the same material. The strength of bonded, or glued, joints can be as strong as the components so joined, particularly in shear. Their weakness is when there is a tendency for the joint to *peel* apart. This happens when there is any bending in the joint. Only in pure tension (or compression) or in pure shear does this not apply. Such situations are not usual unless carefully arranged. A peeling tendency means that the stresses in the joint are concentrated at the peel. The stress increases disproportionately and a crack would start here.

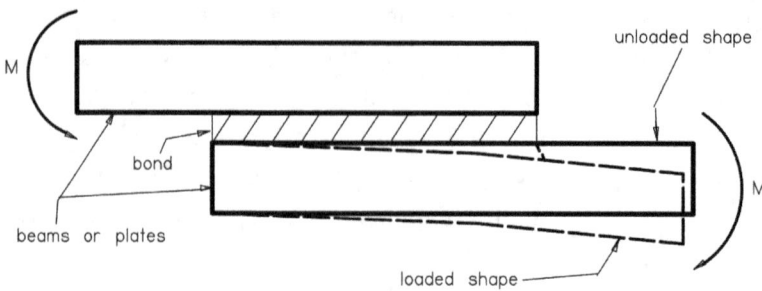

Figure 7.29 Two bonded beams or plates under peel loads, M

Figure 7.29 shows two bonded together beams or plates, although they could be any type of component, being pulled apart by a bending moment, M. The distorted shapes are shown by dashed lines and indicate that the right-hand end of the bond is thicker than the left, although it did not start like that. Obviously, the stress is higher here. The bond failure stress will be reached here first and at a low load level.

The shear stress in a bonded pure shear joint is simply the relevant load divided by the area. Similarly, the tension stress in a pure tension joint is the load divided by the area. Pure shear joints are joints which are in double shear, where there is no bending generated, as in single shear. If, in a tension butt joint, there is not enough cross-sectional area to keep the stress low, a scarf joint – one cut at an angle across the component – can be used. The stress in the bond is then a tension plus a shear, but there is no bending. Joints which have to carry much bending obviously need careful stressing, perhaps by an expert.

There is a potential problem with bonded joints, and that concerns the manufacturing method. It is essential that the surfaces of bonded joints are clean so that the glue sticks to both and not to a layer of dirt or oil or badly adhering paint. Since there is likely to be some poor cleaning, the shear or tension area is reduced, so the stress increases. This is one reason the Safety Factor in the calculation should be increased. Also, the bonding material has to be carefully chosen because it can age or be affected by sunlight, oil, water, and so forth.

Another point to note about bonded joints is that the glue can be brittle. This is particularly true when the bonding material has aged.

Bonds should therefore not be subjected to shock loads. Some adhesives are better, but advice should be sought.

Trapping

Sometimes a component is held by trapping it between two others which themselves are fixed together, perhaps in the manner of a bicycle front fork between the axle and the nuts on each side. There are many other examples. If there is a load on the trapped component, then the stress to be considered is the bearing stress on the contact area. This should not be more than one and a half times the yield stress of the materials if they are metals.

It can be difficult to estimate the contact area if there is any bending involved – that is, if the loading is not uniformly spread over the contacting surfaces. It is best to be cautious if these stresses are approaching the allowable; otherwise, there will be local distortion in the shape of burring of edges or indentations. One approximate way to deal with a problem in which one component bears on another is to assume that the loading between them is of a triangular shape, with the maximum at the contact edge and stretching back a length equal to the thickness of the thinnest component.

Imagine a horizontal slice or section through a door bolt; the fit must be good, a tight sliding fit rather than a loose one. If someone pushes on the door without withdrawing the bolt, a load will be exerted by the bolt on the jamb. The section will look something like Figure 7.30 below.

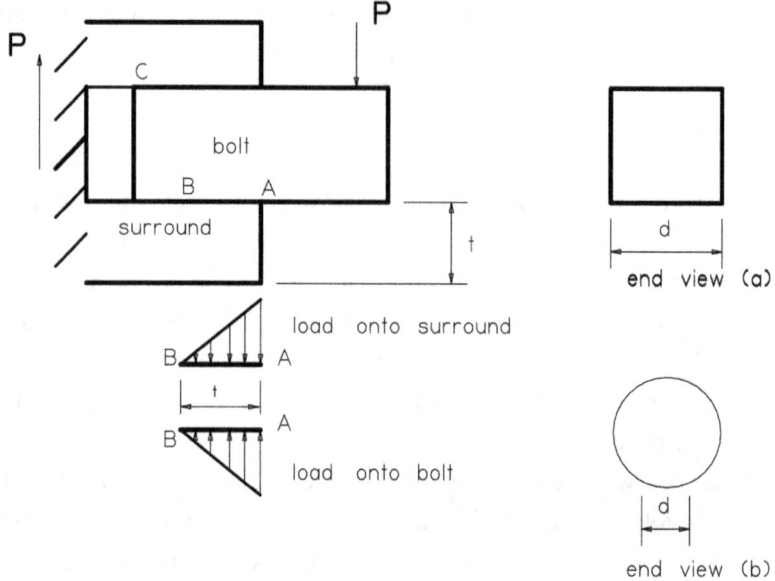

Figure 7.30 Section through a bolt sliding in its surround.

The load, P, is reacted by pressure on its surround. This pressure is assumed to have a peak value at the edge and decrease steadily – that is, linearly – to zero at a distance equal to the thickness of the surround. The peak bearing pressure, w, is worked out from the balance equation by equating the applied load, P, to the average pressure times the area over which it acts.

$$P = \frac{w}{2} \times d \times t$$

$$w = \frac{2P}{dt}$$

P is the load, t is the surround thickness, d is the depth of the bolt in a direction into the paper, and w is the peak pressure on the bolt and surround. In the equation, w is divided by two because that is the average pressure on the area. The bearing pressure is also the bearing stress in the bolt and surround. Obviously, the softer of these two will give way first and show indentation.

The depth of the bolt, *d*, is obvious if it is flat over its contact area, as in end view (a); if it is a round bolt, as in end view (b), however, the length to take is more difficult to ascertain. As the bolt is a tight fit, use 70 per cent of the diameter as a first estimate for d.

If the bolt diameter is smaller than the surround (that is, the fit is loose), then initially the contact width and depth are vanishingly small. There will be no triangular pressure but a point load at the edge at A and also at C. Initially the stress will be theoretically infinite. This cannot happen, so the material will give until d and t are large enough to reduce the stress to the yield value in bearing, about 1.5 x the yield stress of the material.

Friction

This type of joint transfers loads from one component to another by the friction between them. This friction must be produced by holding the components tightly together. This is usually achieved by tightening a screw or wedging things together. It is an unsatisfactory and dangerous method because any sudden load or shock to the assembly tends to break the friction and cause the components to slip past each other. An example occurs on cheap bicycles. Often the brake levers and gear change lever are held on to the handlebars by tightening a screw at the open end of a strap which carries the lever. But it can be difficult to tighten the screw enough without stripping the thread or destroying the screw head. This is obviously an inappropriate design.

The method can be made safe but great care must be taken.

Lugs

These are single pin fasteners in which one inner piece of material (or lug) slides between two others, the pin going through all three. They can usually be dismantled. They may rotate about the pin in service. The pin must be smooth – not threaded – where it passes through the lugs.

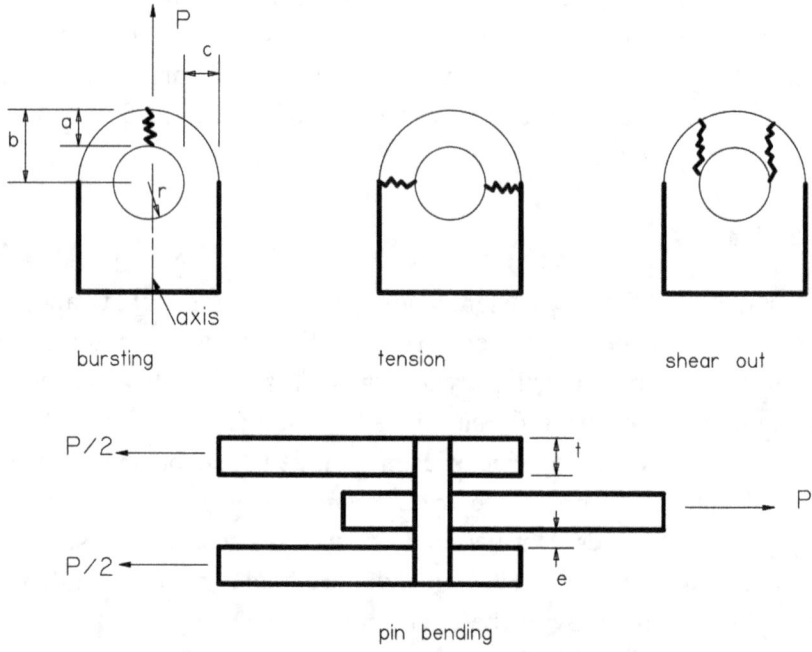

| bursting | tension | shear out |

pin bending

Figure 7.31 Typical lug showing some possible fractures and a section through an assembly of two outer and an inner lug with the pin.

Often the load in all the lugs is tension along the axis of the lug, but if one or both is fixed, the load can have transverse components. The figure shows possible fractures when the load is along the axis. If there is a significant transverse component, its stresses must be added to those due to the longitudinal load. The stress in the lug around the hole is tension and shear plus bending. The last is due to the fact that the pin does not bear completely on the hole surface, so the loading on the hole is only partial. The pin need only be a little smaller than the hole for the load to tend towards being a point load. This obviously elongates the hole, causing the bending. The only exception to this is when the pin is a force fit; that is, the hole is actually stretched by the pin. This causes an extra stress, of course.

There are five types of failure, as illustrated in Figure 7.31. The equations for stress given here are empirical, based on experimental modification of basic equations. This is one way of dealing with complex situations. You write down a simple equation and calculate a stress or deflection, and then you measure this quantity and divide it by the calculated one. If this ratio is constant over a range of loads and dimensions, you have found a *factor of proportionality* or some equally grand name. Curve fitting or a fudge factor we used to call it.

The first failure type is commonly known as *bursting* because the lug splits apart along the fracture shown in the figure. A simple, conservative estimate of the bursting stress is:

$$\sigma = 0.7 \times \frac{P}{at}$$

P is the load along the lug axis, a the throat depth, t the lug thickness, and σ the tensile stress. This equation gives a stress to be compared with the allowable stress but does not necessarily give an actual uniform stress at the throat. Experience has shown, however, that a safe situation can be achieved.

The second failure type is tension on a section through the hole centre line at right angles to the lug axis (see the figure). Again, a simple, conservative equation for the tensile stress is:

$$\sigma = 1.25 \times \frac{P}{2ct}$$

Therein, c is the distance from the hole edge to the lug side, and the other symbols are as before.

The third failure type is due to shear out on two sections, as shown in the figure. Here a piece of the lug pulls out of the parent component. Because the exact length of the fracture is not easily predictable, calculating the stress is difficult. It is between "a" and "b" in the figure. A safe equation is:

$$\tau = \frac{P}{2at}$$

where τ is the shear stress, and the other symbols are as before.

The fourth failure type is the bearing stress of the bolt in the lug. This is the same as for bolts in other situations and has been described.

Obviously, the central lug experiences twice the load that the outer two do, so all the stresses in it are twice as large. Alternatively, the outer lugs can be made thinner and/or sometimes of a different material.

The last type of failure concerns the pin. The failure is as a beam, but there is often a complication. So far in this book, the stresses in beams have found by calculating the bending stresses, which $= M^y/_I$. Remember, though, that one of the conditions of the bending theory was that the beam should be long compared to its depth. This is often not the case for lug pins. The first thing to do, then, is to find this ratio.

The depth of the beam is simply the diameter of the pin. The active length is the thickness of the central lug plus any gap(s), "e" in the figure, plus part of the thickness of the outer lugs. This is because the edge of the hole gives a little under the load. It is sensible to take about one-sixth of the lug thickness on each side, provided the lug is not much thicker than the pin diameter, say, twice at most. If the ratio is at least five, then bending predominates. The pin can be stressed as a beam, and the result will be conservative.

If the ratio is small, which means the diameter of the pin is large compared to the lug thickness, then the load is transferred predominately by shear and the bending stresses are small. The shear stress should be calculated and considered to be most important if the ratio is two or less. If the ratio is between two and five, and the stress worryingly high, advice should be sought. The load divides between the two mechanisms, bending and shear, in proportion to their stiffness.

This analysis assumes the stresses stay elastic and that the maximum stress is the yield stress of the material. For metals, this may be pessimistic. Especially for relatively large pins, it is permissible for a little excessive bending stress to exist.

The ultimate failure strength is likely to be higher but difficult to ascertain. As usual, the advice is to seek expert opinion if the pin in your project appears to be under strength.

7. Stress concentrations

So far, we have discussed stresses which vary smoothly and relatively slowly over space, if they vary at all – that is to say that there are no sudden discontinuities in value. This is because the components have constant or only slowly varying dimensions; plate thicknesses are constant, beam depths do not vary suddenly, widths are constant, and so on. There are many occasions, however, when this is not convenient for the function of the structure. There is a need to drill a hole, introduce a notch, or change diameter or thickness suddenly, as illustrated in the examples in Figure 7.32 below which show a few diagrammatic instances of such occurrences.

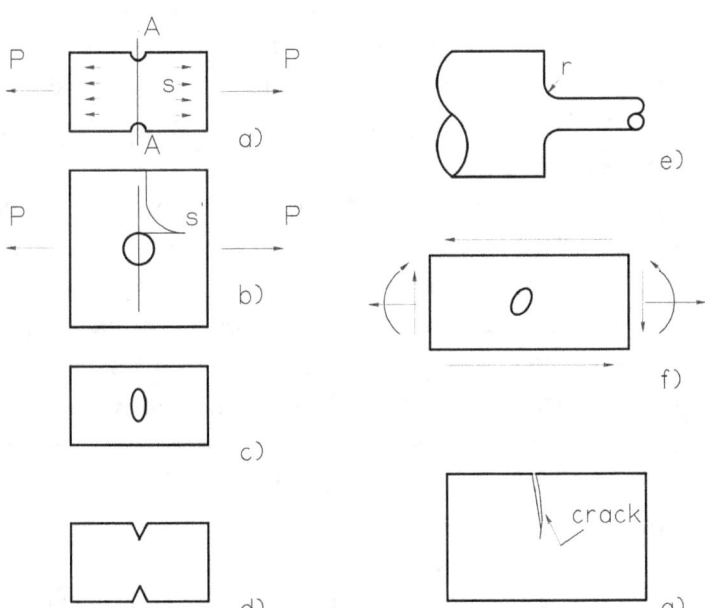

Figure 7.32 Various examples of stress concentrations including a possibly advancing crack

The first instance, Figure 7.32a, shows a plate or a beam with two semicircular notches opposite each other under an applied load P. The stress, s, is constant across the piece away from the notches. At the notches, at the minimum section AA, the stress must be higher because there is less cross-sectional area, so P/A is higher. But this is not the only reason for a higher stress.

If the plate is assumed to be made up of many fibres, then those which are cut by the notch must do something with the load they carry. This load has to move towards the middle into the continuous section. It does not do this to produce a uniform stress across this area; instead, as it were, it crowds around the notch. This is a little like water in a channel flowing over an obstruction. Near the obstruction, it flows faster than away from it. The stress rises very considerably at the edge of the notch.

Figure 7.32b shows a hole in a bar or plate, another frequently occurring stress raiser. The stress in the component due to the end load, P, is plotted across the centre line of the hole. It will be seen that the peak stress, s, is much higher than the uniform stress away from the hole (the *net stress*), which itself is higher than that ahead or behind the hole. For a relatively small hole, the theoretical *stress concentration factor* (SCF) is three. In other words, the peak stress at the edge of the hole is three times the uniform stress. This decreases immediately as you move away from the edge; the effect is very local to the hole, as it is at the notches.

The value of the SCF varies widely depending on the type and size of the discontinuity, the geometry of the component, and the type of loading. Figure 7.32c shows an elliptical hole, while Figure 7.32d shows V notches with a very small radius at the bottom of the notch. In plates, you may have a notch on only one side; in circular bars, the notch can be a continuous groove. Figure 7.32e shows another example where a smaller diameter bar joins a larger; here it is important to have as large a radius, r, as possible. Figure 7.32f shows the various loading types which can occur in any combination that produce stress concentrations, end load, shear, bending moment, or torque.

There is a well-known book which gives many SCFs that have been derived mathematically or empirically in many circumstances: *Stress*

Concentrations Factors, by Peterson. Many other books give a few common cases.

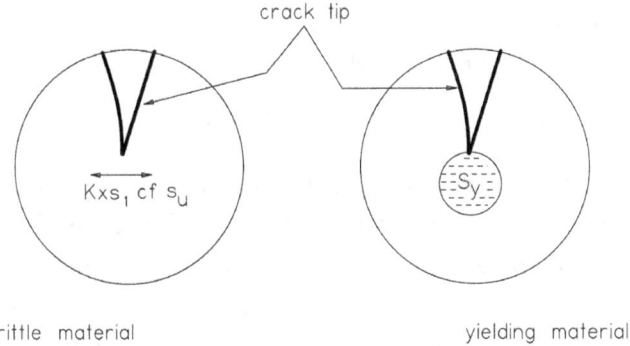

Figure 7.32g, expanded. The effect of different materials on crack behaviour. K is the stress concentration factor (SCF), s_l is the uniform net stress across the plate, and s_u is allowable ultimate stress or the fatigue allowable stress. S_y is the yield stress.

Stress concentrations are especially important in two circumstances, firstly for components made of brittle materials such as most cast metals or ceramics or some reinforced plastics. In these materials, the stress under load increases linearly to failure, with little or no plastic phase, and fracture occurs with very little distortion, only that due to elastic behaviour. The stress concentration therefore reaches the ultimate failure stress, while the general stress is still low. This situation is shown in the left-hand diagram of Figure 7.32g, expanded. This Figure shows the circles around the crack tip.

The second important circumstance, when the crack tip stress is still elastic, is when the component suffers a repetitive load; the failure could then be by fatigue. The peak concentrated stress is above the fatigue allowable stress. In both cases, the nominal stress is still elastic.

When, as in the right-hand diagram, the material is a metal which has a plastic phase, the situation is different. At the yield point at the tip of the stress concentration, the material yields and deforms plastically; that is, the material stretches at a faster rate than if it were still elastic, so it will not carry as much load as it would if still elastic. The material next to this region, which is not stretching as fast, has to carry the excess instead. Because of the localised nature of stress concentrations, this excess load is not very large so the adjacent material can carry it. The situation, then, is that the stress at the edge of the concentration is a little above yield in the plastic region, but not at fracture stress, and the adjacent material is at a slightly higher stress than expected, but not seriously so. The component stabilizes. There may be no further plastic deflection.

Stress concentrations can be caused initially by very small blemishes in the surface of a component. These might be caused inadvertently during use or manufacture. If they have sharp tips at the bottom, which in turn implies a large SCF, a relatively low working stress will then be multiplied to near failure. A crack develops. Under repetitive loading, such cracks may at first, slowly and then faster and faster, grow to catastrophic failure. This matter is a large subject known as Fracture Mechanics and will not be discussed here.

8. Poisson's ratio, v

If a bar is pulled into tension, then it shrinks in cross section; if it is compressed, it expands in cross section. For *isotropic* elastic solids, the shrinkage/expansion is proportional to the load-induced elongation or compression. This finding is due to Poisson, a French mathematician (1781–1840), and is known as Poisson's ratio. Isotropic means that the properties of the material are the same in all directions. So if there is a strain in the x-direction – say, along the bar – then there are strains of $-v$ times that strain in the x-direction in both the y- and the z-directions, where the y-direction is the up and down direction and z-direction is sideways. If ε is the strain, then in equation form:

- In the x-direction, the strain is ε_x
- In the y-direction, the strain is $-v\,\varepsilon_x$
- In the z-direction, the strain is $-v\,\varepsilon_x$

There is a reduction in the area of the cross section of $(1 - v\,\varepsilon_x)^2$. While this is interesting, at the level of this book, it does not often impinge on the matter. However, some equations use the ratio, so it must be mentioned. The mathematics shows that for these ideal conditions (isotropic elastic), v is 0.25. For most steels, it is 0.33; for aluminium, it is 0.31. This is a usual value for most metals. It is not too important, as it does not usually make much difference to the answer in simple equations. For materials which do not change their volume under stress – for instance, rubber – the value is 0.5, which is the maximum possible.

Section (d) Finite Elements

It has been mentioned frequently that if you take recourse to experts, they may reach for their computers and use a type of program called Finite Elements Analysis. This is basically a mathematical representation of all or part of a structure. It is done by depicting the structure as being made up of a continuous series of small elemental divisions of itself. The corners of the meeting points of the elements are called nodes. The act of dividing the structure into these elements and nodes is called meshing the structure. Figure 7.34 is a bar with a hole which has been meshed. Figure 7.33 shows typical elements. The little circles depict where the nodes are.

The elements meet at the nodes (but not necessarily along their sides), and here the mathematical equations are generated. Without going into further detail, the equations contain the deflections as unknowns. Many simultaneous equations are generated. The advantage that computers have over people is that they can solve any number of these equations, millions nowadays, depending only on the size and speed of the machine. People cannot reliably solve more than about a dozen, and they take months to get it right. At school, if you remember, they never gave you more than two or three. A former colleague, working entirely with pencil and paper before computers became available, once spent three months getting a correct solution to a set of twelve simultaneous equations. He went on holiday to celebrate. Nevil Shute, the novelist and an early stressman, tells of an almost religious experience when successfully solving such problems after months of work on them. Praise be to computers.

Besides these deflection equations, the normal balance conditions and the constraints on the structure are incorporated into the calculations.

The results give deflections, loads, and stresses at every node, and these can be listed (a very tedious way of examining answers) or plotted on a picture of the structure as contours which can be assigned different colours, so-called fringe plots. So a red band of colour may indicate a high tensile stress, while a blue one indicates a high compressive stress, with other colours showing intermediate values. Pictures of the structure showing highly exaggerated deflections are also generated. All this gives a much more comprehensive and detailed picture of the effect of a set of loads on a structure. Doing the calculations previously described in this chapter, by contrast, only gives peak values of the deflections and stresses. The location of these peaks is left to the experience of the stressman. Today's professionals would use this as a check on their FE, while years ago it had to do the entire job of supporting the design process for strength assessments.

There are many FE suites available, mostly for money. In the eighties, a large engineering company undertook a survey with a view to buying a suitable version. They considered four hundred suites. Since then, new ones have arisen and old ones become defunct. Many individuals write their own. Some are specialised, and some, most probably, are general and claim to be able to tackle any sort of problem. Because they all solve the analytical problem at discrete points only, they all give approximate answers. However, the error can be reduced to small proportions by using many nodes and by tying these together with better elements.

In general, all providers of FE use similar sets of elements. Historically, as simple as possible elements were used because computers had little memory and were slow by the standards of today. Figure 7.33 below shows how element formulations progressed.

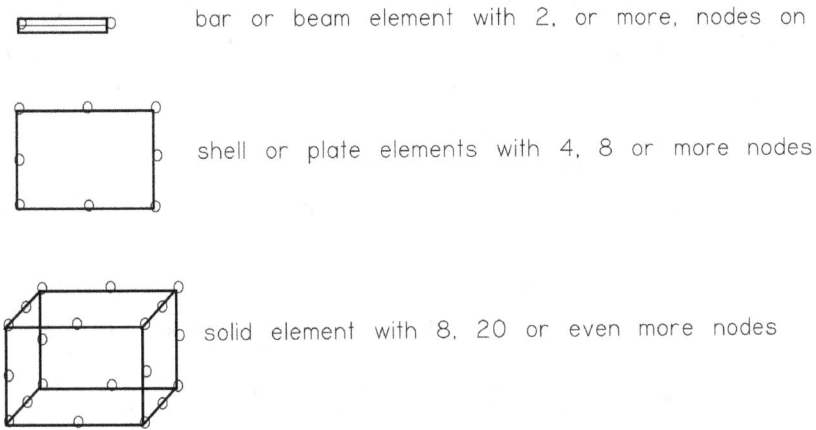

bar or beam element with 2, or more, nodes on

shell or plate elements with 4, 8 or more nodes

solid element with 8, 20 or even more nodes

Figure 7.33 Typical elements and nodes

Originally, simple bar elements were developed. These could carry only end loads, tension, or compression, and they were used to analyse complex frameworks. Later, bending capability was incorporated which extended their usefulness, now known as beam elements. These elements are mathematically exact – they give the same answers as the classical methods. Each element has its cross-sectional details individually defined in the FE model so beams of varying section can easily be specified. The beams do not have to be straight either.

Next to be developed were shell elements. At first, these also could resist only end loads and had nodes at only the corner points. Later, like beams, they were developed to carry bending moments. Only having corner nodes means they only necessarily touch at the nodes. The deflections in between could only be linear interpolations of the corner nodes. By inserting mid-side nodes, the deflections of the sides could be depicted as curves, which is more realistic. The elements still only touch at the nodes, but the inter-node discrepancy is smaller.

Similarly, solid elements with eight and then twenty or more nodes were introduced. The reason for this slow development was that computers had to increase their capacity to cope with the ever-increasing demand. At each node, there are what are known as degrees of freedom, anything from one to six. Each linear motion has a degree of freedom and so does each rotation. Bar elements have one degree at each end; beams can have all six. Each degree means one equation.

Clearly the number of equations can and do skyrocket. In the old days, the 1970s, the computers were mainframes but the models were still quite small and took hours or even days to run. Nowadays, desktops run such models in a minute or two and can manage much larger ones.

So the first job when analysing a structure by FE is to mesh it. All points of interest are given nodes so that the answers can be read from the output results directly – that is, changes in geometry, points of load application, points of expected maximum stress or deflection, and so on. Figure 7.24 shows a simple bar with a hole; the bar is in tension.

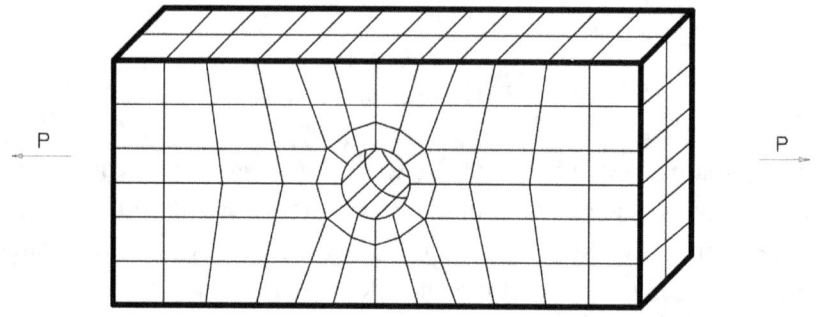

Figure 7.34 Meshed bar in tension with hole

Notice that although the ideal shape for a solid element has all its angles equal to right angles, and thus to be cubic, this cannot be achieved in a practical mesh where circular shapes occur. It is necessary to try to make all the angles as near a right angle as possible; certainly very acute angles give poor answers. If the purpose of the analysis is to find the peak stress due to the hole, then this mesh would need to be much refined. The elements near the hole would be split into many smaller ones. Alternatively, if the idea is to find the lengthening of the bar due to the loads, P, the above, coarse mesh would do. Deflections are more accurate in a given mesh than stresses due to the mathematical methods used in most FE suites.

The bar could have been represented by a model of only a quarter of that in the figure – say, the top right portion. This is because it is doubly symmetric about the horizontal and vertical centre lines. Special constraints have to be applied at the faces of symmetry to recognise this. The answers are the same.

Much meshing can now be done automatically by computers which saves most of the time that used to be spent setting up a FE calculation. This means that the strength calculation phase of any project can now be performed at the design stage and the process used to optimise the structure if this is important for weight or financial reasons. The process was formerly so time consuming that FE was used only to check that a structure was strong enough. The discovery that it was not and that the structure had to be strengthened was expensive, as any retrospective corrective work always is. It is, of course, much more expensive to learn that your structure is not strong enough when it is manufactured and in use.

When a FE calculation has been done, it must be checked for accuracy. There are many types of checks which I will not go into here, but never believe the answer without the checks.

Section (e) Dynamics

In this section, we consider any loading situation that is not a constant or static one. They may not be large in themselves but can have damaging effects if their frequency coincides with the natural frequency of the structure they are applied to. These loadings have been described in chapter 1, section (c). Their reactions are always dependent on the natural frequencies of the structure. All but the simplest structures have many natural frequencies; each has a different deflection shape, usually called a mode. The lowest frequency, the fundamental or first mode, has the largest and simplest deflection. It is usually but not always the most important mode. Below are a few simple equations to find the natural frequencies of simple components; this may alert you to a potential problem.

Mathematicians talk about *degrees of freedom* (DOF) of the structure or systems generally – for instance, swinging pendulums, organ pipes, hydraulic arrangements, and so on. Simple systems have a single DOF, and this is defined as having just one displacement, which leads to comparatively simple equations which they can (and have) easily solved for all sorts of basic structural components. There is a book rather like Roark, written by Blevins, which collects all these simpler formulas to tell you what the natural frequencies of beams, plates, and so forth, are, plus some other vibrating systems. It is quite mathematical, but the tables can easily be used as much as Roark. They also give higher modes for simple structures. As usual, the structures are mostly of uniform section and assumptions about support conditions must to be made.

Beam

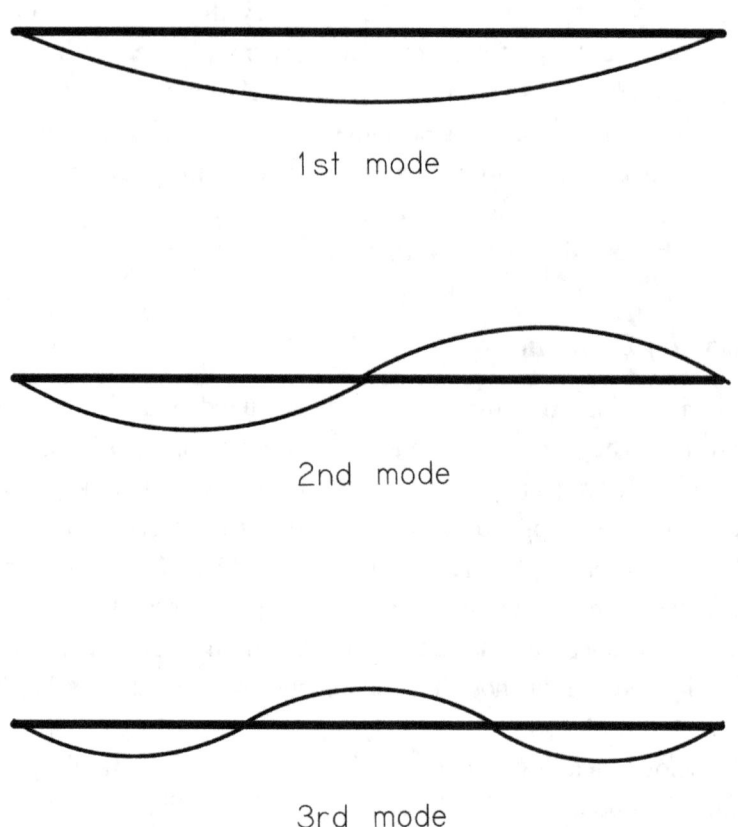

1st mode

2nd mode

3rd mode

Figure 7.35 A simply supported beam showing the first three vibration modes

The SS beam in Figure 7.35 has been disturbed from its rest position, which is straight. It is of uniform section and made of a material with linear stress/strain characteristics. Mostly it will vibrate in its fundamental mode. Only in special circumstances which force it to will the beam vibrate in the higher modes. The mathematics shows that the frequency of vibration is proportional to the inverse of the length squared, $1/L^2$, where L is the length; so the longer the beam, the more flexible it is and therefore the slower it will vibrate.

Just by inspecting the shape of the modes, Figure 7.32 suggests that for the second mode, you could substitute ½L for L. In other words, it is four times faster than the first mode. Experience confirms this. The third mode will be nine times faster. These sorts of estimations sometimes work; you have to examine the deflected shapes and see whether there is an obvious ratio to the lengths. In the figure, the second mode obviously divides the beam's length into two; the third divides the length into three.

A simple equation, applying to single DOF structures and where the deflection under a static load has been calculated, is as follows:

$$ f = \frac{1}{2\pi} \sqrt{\frac{g}{d}} $$

Herein, f is the frequency which is measured in cycles per second, or Hertz and the motion is in the direction of the load. A cycle is completed when the motion goes through the starting position again, i.e. if a point on the structure starts at its maximum deflection, say up, travels to the maximum down, and back to the maximum up, it has completed one cycle. Also d is the static deflection, g the acceleration due to gravity (9.81 m/s), and π is the usual. This gives an estimate of the frequency to within a few per cent. It should give an idea of the value of the fundamental vibration frequency of an individual component within the structure if you have a problem.

Membranes

A membrane is a plate which is relatively so thin that its bending strength is minimal. If it is stretched into tension in all in-plane directions, it will vibrate if excited, without any significant bending stresses. A drum skin is an obvious example. The fundamental mode, or distorted shape, is the membrane assuming a dome shape. Higher modes can have several waves in the skin, all centred on the centre of the circle, or they may divide the distortion of the skin into halves along a diameter, or two diameters at right angles, or some combination of any of these and more. In the case of the drum, the higher modes provide the overtones. An equation for a circular membrane is:

$$f = 0.679 \sqrt{\left(\frac{S}{\gamma A}\right)}$$

Here, f is the frequency in Hertz, S is the tension per unit length of the circumference, γ is the mass per unit area of the membrane, and A is the area of the membrane.

It will be noticed that the membrane must be tensioned, which any drummer will recognize. Also the density of the skin is involved (the heavier, the lower the frequency) and the bigger the drum, the lower the tone; all is as expected.

It is a fact that any membrane of a shape which is not too dissimilar from a circle has a similar fundamental natural frequency. So a square, a quadrant, or a sector of a circle or semicircle, an equilateral triangle, or even a 2:1 rectangle, has a fundamental not more than one-sixth larger than a circle of the same area. All this gives a useful first estimate of the vibration that is likely to occur.

Plates

In contrast to a membrane, plates principally bend. This may be because they are thicker, or because the edges are not supported against an in-plane tension force, or because there is no such force. Again, there are equations you can use to find out whether your gismo is vibrating badly due to some unfortunate provocation. Blevins gives them. The first thing to do is to calculate a characteristic called the *flexural rigidity.* This defines your plate.

$$D = \sqrt{\frac{Et^3}{12(1 - v^2)}}$$

D is the flexural rigidity, E is Young's modulus, t the plate thickness, and v a property of materials called *Poisson's ratio*, which has been touched upon in Chapter 7 section 8.

Now, to get an idea of the fundamental natural frequency of various plates we can use the following equation:

$$f = \frac{\alpha}{a^2} \sqrt{\frac{D}{\rho t}}$$

f is the frequency; α is a number, given below, which describes the shape of the plate and its support; a is the length of plate or its radius; D is given above; ρ is the density or the mass per unit volume, kilograms per cubic metre in metric measurements; t is the plate thickness. A few simple examples of α are given below. These arise out of the theory and sometimes practical experiments which underlie this subject. They give reasonably accurate results.

Square plate, free edges	$\alpha = 2.147$
Square plate, simply supported edges	$\alpha = 3.142$
Square plate, fixed edges	$\alpha = 5.728$
Circular plate, free edges	$\alpha = 0.836$
Circular plate, simply supported edges	$\alpha = 0.792$
Circular plate, fixed edges	$\alpha = 1.627$

It will be noted that for circular plates, the SS case has a lower α and therefore a lower fundamental than the free case. This is not as expected and not true for square plates. It has to do with the deflected shape of the plate. The SS case assumes a simple dome shape, but the free case does not. Its lowest mode bends about two axes at right angles to each other.

Rectangular plates are dealt with in Blevins's book using a similar formula but with different values of α, depending on the length to width ratio.

If you know the central static deflection of your plate under its own weight, you can use the simple formula for the fundamental mode, with 1.227 substituted for the 1. This applies to variously shaped plates with various boundary conditions and gives a rough and quick idea of the frequency of vibration.

The above situations apply to the individual parts – the beams, membranes, or plates – which are usually only a part of a bigger structure. This structure has an influence on the frequencies because it provides the boundary conditions for the part under consideration; that is, it restrains the part as SS, or fixed, or in between. Also, it will have its own natural frequency, which may contribute to or change the frequency of the individual part. For these reasons, the answer given by the simple equations above are only approximate. If a more accurate answer is required, then the effect of the bigger structure will have to be taken into account. This needs a specialist who will probably use a Finite Element program.

Vibrations will die away unless they are continuously provoked. This is because there is always something called *damping* present. It may be so light that the vibrations go on for a long time; it may be so heavy that it restricts the vibration to a single cycle. For instance, a bell struck once will ring for some time; but a car wheel should have a damper so strong that a bump in the road, causing the wheel to deflect up, will immediately be returned to its proper position. The bell will lose energy because it is causing the vibrations in the air, and to a lesser extent, there will be internal friction in the material, both constituting damping. The wheel damper acts by forcing oil through a small hole from one chamber to another within the damper. There are many sorts of damping and complicated ways of assessing their influence. Again, this is outside the scope of this book; consult the experts.

Section (f) Summary

In this chapter, we got down to cases. After listing the basic assumptions, both physical and mathematical (see the contents for Chapter 7, section (a)), units and their relationship to each other are covered; relevant SI units are listed and some conversion factors given.

The various types of stress are explained, pure tension or compression, bending, shear, and torsion as well as how to combine them. An important section on various methods of joining parts comes next.

Two more esoteric subjects are mentioned for the sake of completeness. These are stress concentrations and Poisson's ratio. A section on the Finite Element method is included because that is what professionals mostly use these days. Finally, the dynamic behaviour of structures is described to a simple level and to draw attention to the fact that here is a whole subject that cannot be ignored.

STEP FOUR

CHAPTER 8

Stress and strength comparison

This is the last part of the process of stressing a structure. The loads have been calculated for all the cases and the largest picked; the safety factors have been chosen and the loads increased by them and the resulting stresses calculated. Now it has to be decided how high these stresses can be allowed to rise. This depends on the material and the nature of the load case. First the various classes of materials will be discussed, then their allowable stresses according to the service conditions, and finally how to calculate the *reserve factor*.

Section (a) Materials

To be a stressman, it is not necessary to be a metallurgist or materials scientist, but it is advisable to know a few things about materials. When designing a new structure, it may be required to choose a material for it. The criteria to be considered will be strength, obviously, but also operating temperature and the effect of contact with other media such as liquids and gases; they may corrode the structure. In addition, there will be financial and manufacturing considerations. Great care has to be taken to discover the exact specification of the materials of an existing structure if you are checking on its use or thinking about changing the use.

The first thing you need to know about the material of a structure is whether it is brittle or ductile and, if the latter, to what extent in terms of strain to failure. The second is whether it behaves linearly over at least part of its stress-strain curve. If the material is brittle, then it will eventually fail suddenly with no warning signs. You therefore have to be more careful about the stressing, making sure the stresses are low and accurately known. If there is a ductile or plastic phase after the elastic one, you have an extra margin of safety. This is because the component will stretch and bend before fracture, which may or may not affect its performance, but it may not be catastrophic.

If it has a linear phase in its stress-strain curve, then normal stressing can be applied. If not, special and more difficult calculations have to be used, nowadays probably non-linear FE.

Consider each type of material individually.

1. Metals

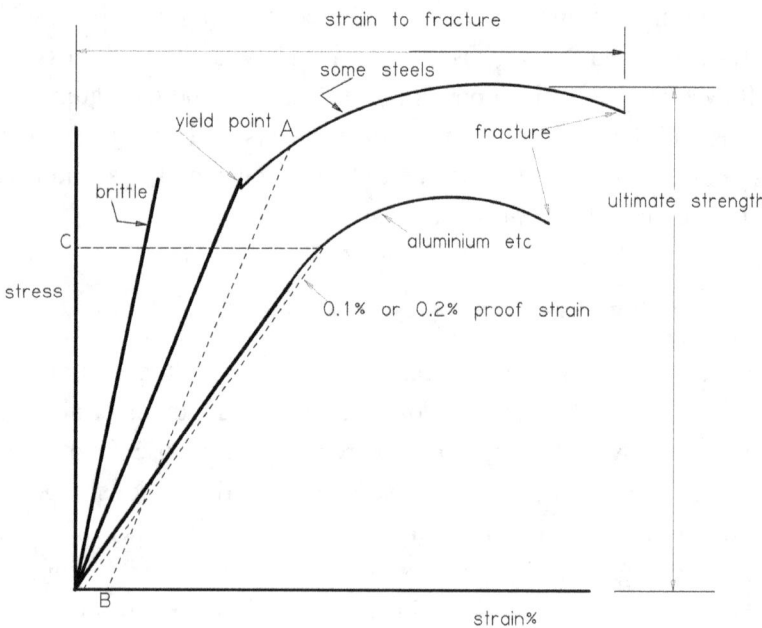

Figure 8.1 Typical stress/strain curves for metals

The axes of the graph in Figure 8.1 are stress and strain. Stress in a component of the structure is proportional to the load on the structure.

Strain equals the change in a dimension of the member due to a load divided by that dimension when there is no load. It is proportional to the deflection of the structure. The figure is similar to Figure 7.1.

The curve marked "some steels" in Figure 8.1 is straight up to a small drop at the yield point, after which it is curved up to the end, which is the point of fracture. The stress rises up to a maximum, called the ultimate strength, and then falls to the point of fracture. This fall is due to severe metal flow just before fracture. In a bar under tension, for instance, the bar diameter reduces severely; this is called necking. The yield point shown applies only to steels up to a middling strength. So up to the yield point, the curve is linear and normal calculations apply. If the load is removed, the structure returns to its original size and shape. After the yield point, the increase in strain is higher for a given increase in stress. If the load is removed from the stress/strain situation at point A, the member's stress/strain path is along line AB, which is parallel to the original linear curve. It stops at B, which shows that there is now a permanent strain in the structure. For instance, if you take a straight steel bar and bend it only a little, it returns to straight when you remove the load. If you continue bending, it remains bent on load removal. Because most structures should not gain a permanent set in normal use, they are designed never to exceed the yield stress. Notice, however, that the component does not necessarily fracture just because it has exceeded the yield stress by a bit. The distortion in the structure may not even be noticeable. It may not matter much. Metals, if they are not brittle, are forgiving in this way.

The final fracture for low strength steel could be at 30 per cent strain, which implies a large distortion of the structure. This is put to use in cars where the force from a frontal collision is absorbed by high strain crumpling of the front of the car. This reduces the severity of the crash.

Steels of high ultimate strength have reduced strain to failure; the higher the ultimate, the lower is this strain. This means the material changes from a ductile to a more brittle state. This has implications in design, as brittle structures do not sustain impacts as well as ductile ones. This seems to be true for all types of material. There is a balance between brittleness and high strength. The line marked "brittle" on the graph makes the point.

Aluminium, like other structural metals such as titanium, high-strength steels, magnesium, copper, and their alloys does not exhibit a marked yield point. Instead, a stress called the 0.1 per cent proof stress is used as the allowable stress. This leaves a 0.1 per cent strain when the load is removed from a component. The figure might be 0.2 per cent. With a Safety Factor of at least 1.25 on the loads, this is not a significant point, as no permanent set would ever be expected. The value is found by testing samples of the metal and drawing curves such as the one in Figure 8.1 and constructing the marked line parallel to the linear part of the stress/strain curve such that the intercept on the strain axis is 0.1 per cent or 0.2 per cent. The intercept on the stress axis at C then gives the allowable stress required.

Metals are usually treated to develop their strengths. This involves adding various other elements, heat treatment or mechanical working of some description, or some combination of these. Mechanical working involves hammering, rolling, or extruding the basic cast material. This refines the grain structure of the metal. It may also produce directional strength properties. If a sheet of metal is rolled out, then the grains are elongated in the direction of the rolling, and this gives a better strength in this direction; this is the long direction. At right angles to the rolling direction (known as the long transverse direction), the strengths are less. The through the thickness direction (short transverse direction) has the least strength. This phenomenon is especially evident in aluminium alloys and can be marked – say, 15 per cent – so if there is a possibility of it being critical, advice must be sought from a metallurgist. It will be important when a major stress is acting in the *transverse* direction – that is, at right angles to the rolling or forging or extruding direction. Quite detailed knowledge of the manufacturing method is required.

Castings which have not been treated are brittle. They fail suddenly, with little or no ductile phase. This is because they have sharp-edged faults and inclusions that exhibit high stress concentrations. So although the overall stress in the cast component is quite low, there are areas throughout which have very high stresses causing failure. Castings can sometimes be treated to avoid this by a process known as Hot Isostatic Pressing (also known as HIPping). The cast components are put into an oil-filled pressure vessel and heated. Then pressure is applied. The faults and inclusions are squashed. Presumably, some working of the material also occurs.

2. Reinforced Plastics or composite material

These often replace metals or timber nowadays. In their place, they have superior properties in that, among other things, they are easily made into nearly any desired shape, can have a large range of strengths and stiffness, are corrosion resistant, can take up any colour, are electrically and thermally insulating and are usually lighter than metals.

They are made up of two elements, the reinforcement and the matrix. In general, the reinforcing provides most of the strength and the matrix holds it together. The reinforcing material can be continuous filaments of almost any material but is usually glass or carbon. These are arranged in the component to resist loads along the fibre length. This is the most efficient way to reinforce, but across the fibres, the component strength is much reduced. The fibres can be woven mats having good strength in two directions, less so at forty-five degrees to the fibres. They can be of short length and randomly arranged so that the strength is the same in all directions. These are cheapest but have the least strength. The reinforcing can be particulates or even powder, weaker still, but they can be moulded in special machines and so forth. Concrete is a common example where the aggregate constitutes the reinforcement, the cement the matrix.

The *matrix* is usually some sort of plastic, mostly polyester, also epoxy for better strength or phenolic for fire resistance. There are numerous alternatives for specialist applications. The matrix provides most of the other properties and advantages while also exchanging loads between fibres. Note that steel bar reinforced concrete is also a composite material. More advanced composites can use metals, aluminium, titanium, and magnesium as the matrix, with various ceramic particulates as the reinforcement. Again, this is a specialised procedure and manufacture.

The stress/strain behaviour of the composite is most nearly like the line marked brittle in Figure 8.1, but it changes with the angle of the load to the fibre direction, except for the random fibre lay-up. Only in the latter case can normal, simple stressing be used. However, the properties are much more variable than that of metals. Books and official standards can be consulted, but the manufacturer will have to be involved in the strength specification of particular components. He may not be able to achieve what is wanted with his methods.

There are some highly complicated strength equations for use in analysis and design when there are loads on a component, giving rise to biaxial or triaxial stresses. These are not necessarily applicable to all situations in the sense that the equations so far considered are. If they are based on tests, then they can be found to be accurate only for that test.

If the reinforcement lay-up is at all sophisticated, in that it is specified for particular loads, the procedures required are complex. Specialists will have to be called in.

3. Unreinforced plastics

These are polymers – that is, long chains of organic molecules all intertwined. Their behaviour and strength are dependent on temperature, time, and rate of load application. Although simple stress methods can be used tentatively, the stresses in them should be low compared to their short time strength and the loads should only apply for short periods. They should not be left under load. The author once bought some cable as jump leads for starting cars with flat batteries. There were crocodile clips at each end. These were partly constructed of plastic. Unfortunately, they were so designed that the strong steel springs in them loaded the plastic arms in bending even when not extended on the battery posts. They looked all right in the shop, but after a year of storage in the car, they split and fell apart. They had never been used. Plastic clothes pegs exhibit the same behaviour. This illustrates the phenomenon of creep. So never leave plastic under load for long. Apparently, the long chain molecules slowly slip past each other and either reduce the load or eventually fracture.

The stress-strain curve depends on the temperature. The material can be nearly linear to a brittle fracture, like the curve marked brittle in Figure 8.1, except it is never quite straight. At higher temperatures, plastics can be more like mild steel, and at higher temperatures still; the line marked plastic in Figure 7.1 shows the generally nonlinear relationship between stress and strain. Since simple stressing assumes linearity, it is obvious that it can only be used for the beginning part of the curves. If a few quick calculations show that the stress is small, the answer will not be far wrong, but at apparently high stresses, it is necessary to use more sophisticated methods.

Rubber is also a polymer and has the property that it can non-linearly undergo very large strains and then recover its former shape completely. It is usually used not for its strength properties but its spring-like ones. It would not be designed to work near its fracture strength. It has an analytical methodology of its own. See a specialist.

4. Ceramics

The trouble with ceramics is their extreme brittleness. We all know you cannot drop teacups, for instance. Actually, ceramics are intrinsically strong at the microscopic level. The cause of their brittle nature is that internally, under the surface, they have many small flaws. These are sharp and therefore cause large stress concentrations. The effect of any load is therefore intensified.

There are some manufacturing methods which claim to avoid flaws so the situation may change. These flaws are usually much more serious when the component is in tension so the compression strength is good while the tensile is not. Common examples are bricks and mortar, rocks, and stone.

5. Timber

This is one of the oldest structural materials. It has a bending capability which is better than stone, the other old material. There are a huge number of varieties of trees which are used for special purposes. Structural use in houses, for instance, is controlled by regulatory bodies. Depending on the wood and its condition (moisture content and so forth), these bodies say what the design allowable stresses are. *BS 5268-2:2002* is one such code of practice.

6. Magiconium

This stuff is often hawked around. It is said to be perfect in every way. All the strengths, ultimate, yield, fatigue, creep are better than what you have. It never corrodes and is easy to machine or cast or extrude. And it's cheaper. Beware.

Section (b) Allowable stresses

There are a number of allowable stresses that have to be recognised according to the circumstances and the loading case. These have been alluded to in earlier chapters and will be considered in more detail here. They are generally known as the ultimate strength, the yield or proof strength, buckling strength, the fatigue strength, stress corrosion, and the creep strength. Other load-limiting factors are not strengths but mainly a limit to the distortion of a structure and a vibration limit. We will consider all of these in turn.

1. Ultimate

If a component fails because this strength has been exceeded, a fracture is evident. In tension, a bar, for instance, breaks into two pieces. If the material is a ductile one, the fracture face will be at an angle to the centre line and a neck will have developed – that is, the thickness will have reduced. If the material is brittle, the fracture face is at right angles to the centre line and there is no necking. Obviously, there are stages in between. Plates behave in a similar way. In compression, for a short bar where there is no buckling, the distortion is not necking but a bulging out of the bar followed by fracture with a sloping face. If the material is brittle, it may crumble.

In bending, the fracture obviously starts on the tension side of a bar, and if the load continues to be applied, it extends through the bar until complete separation is achieved. The deflection of the bar at fracture depends on its ductility and the depth-to-length ratio. A shallow, long, ductile one can bend double before a fracture appears. On the other hand, a very brittle bar will snap after little bending deflection.

In service, this should never occur, provided the structure is being used within its design envelope and it has been manufactured according to its design specification, because of the Safety Factor. This is usually set at a minimum of 1.5, so the applied stresses are not near the ultimate strength. The exceptions to this are rare; they apply when other factors make structural integrity irrelevant. For instance, in a severe pullout from a dive in an aircraft there is no point in making the aircraft survive when all the flyers are dead from excessive 'g' forces.

In general, ultimate load cases do not occur often in a component's life – usually only once. Otherwise they become fatigue cases.

2. Yield or proof strength

This strength applies to most metals, timber, reinforced plastics, and ceramics. Up to this point on the stress/strain curve, the deflection in a component or structure is proportional to the load on it; when the load is removed, the deflection also returns to zero. The structure is as it was before the load was applied. This point has been made before.

In most circumstances, this is an important design point. Normal loading experienced when a structure undergoes its duties will not permanently affect it so it can be used again in exactly the same way. This avoidance of permanent distortion is important in any circumstance where the shape of the structure is necessary for its performance. For instance, an aeroplane wing develops its lift because of its curved surfaces, so they must be maintained for it to sustain its function. In many industries, the yield strength is the basis of the design process, the ultimate being ignored. This is sensible, as the artefacts need to be used often and should not be bent or deformed, much less fractured. There is, of course, a small deflection in the structure at yield, but this is only a few parts in a thousand, and this is usually, but not always, unimportant.

As mentioned, if there is no proper yielding in the material, the stress at 0.1 per cent or 0.2 per cent strain is used.

The allowable stresses given in material data books are derived from tests on each type of material and its alloys. Hundreds of special test pieces are stretched to fracture and the stress/strain graphs drawn. Statistical analysis then gives yield or proof strengths with given probability values. When material is later produced to the same specification, only a few tests are required to establish its consistency with the book value. All this is laid down by the relevant authorities.

3. Buckling strength

This strength applies to components in compression and is a function of the material, the geometry of the component, and the nature of the support. It therefore has to be calculated for every component in a structure that experiences a compression stress in any load case. Any type of component can buckle – that is, bars, plates, tubes, or complete structures such as the lattice towers supporting tower cranes for buildings. The last is known as an overall buckling mode; if a single component of a structure buckles, it is called a local mode. Components of complete structures can buckle when the structure is in bending or in torsion or a combination of them. Plates can also buckle under shear loads. Bars usually buckle first into a single curve but in some circumstances may adopt a shape consisting of several waves between supports. Plates supported all round their periphery buckle into quilts (waves in two directions); similarly with tubes.

There are formulae for calculating the buckling stress of many types of components in many situations. These can be found in the usual books, such as *Roark* and *Machinery's Handbook*. As usual, the first thing to do is to classify the component into bars or plates or cylinders or complete structures. Then look at the support conditions, fixed or simply supported or free; each end could have different conditions.

The simplest equation for bars is due to Euler (1707–1783), the famous mathematician:

$$P_{cr} = \frac{\pi^2 EI}{l^2}$$

P_{cr} is the critical buckling load for the bar whose ends are held laterally but can rotate freely, E is Young's Modulus, I the least second moment of area, and l the length of the bar. We choose the least I if the cross section is not symmetrical, that is a circle or square, which have only one *I*.

If the bar or column is relatively long, this is acceptable; otherwise, different formulae have to be used. Long or short is decided by calculating the *slenderness ratio.*

$$\text{slenderness ratio} = \frac{l}{r}$$

Herein, *l* is the length of the column and *r* is the least *radius of gyration.* The latter is a concept to do with the cross-sectional properties of the column. It is defined as follows:

$$\sqrt{\frac{I}{A}}$$

I is the least second moment of area, as defined in chapter 7, section a) (9), and *A* is the area. If this quantity is greater than about 100, count the column as long; otherwise, it is short.

Now choose the relevant end conditions and you should get a good idea when the column will buckle. The Euler equation is

$$\sigma_{crit} = \frac{k\pi^2 E}{(L/r)^2}$$

Therein, σ_{crit} is the stress at which the column buckles, *k* is a factor defining the end constraints of the column, *π* is the well-known ratio between the area and circumference of a circle which pops up all over, *E* is Young's Modulus, *L* is the length of the column, and *r* is the least radius of gyration, as above.

The three equations above can be combined and manipulated to give this:

$$\sigma_{crit} = \frac{k\pi^2 EI}{AL^2}$$

185

Now, $k = 1$ for a column that is simply supported at both ends (the ends can rotate at will but do not move sideways). This is probably the most conservative assumption for components of frameworks. Also $k = 0.5$ if one end is fixed and the other free of all constraint. Again, this is a conservative assumption for any structure, such as a crane tower pointing straight up. There are other values, given in the books, for other conditions. However, care must be taken when assuming rotational fixity, as this is difficult to achieve. For a fixed – fixed column, k is mathematically 4 but in practice should be assumed to be less, probably not more than 2.5. If the critical stress is less than the yield, then you have found a limit which must be observed. If it is higher, it can be ignored and the yield used.

One important qualification concerns eccentricity of loading, which leads to uncertainty. No real components are loaded exactly along their axes; a load is either slightly to one side of the neutral axis or it is at a slight angle to the axis. This results in a moment being applied to the component, which will bend it and reduce the buckling strength.

If the buckling stress is less than the yield on removal of the load, the buckles will disappear and no damage is done. This happens in *redundant* structures. These are structures with multiple load paths, so if one component fails, others can successfully carry the load. In this case, the failed component would be the buckled one. An example is the top skin of a well-known though now superseded airliner; in gusty conditions, this was allowed to buckle, the BM being resisted by the wing spars.

So far, no definite allowable stress has been defined because buckling is not as predictable as simple tension or bending. Many empirical equations have been developed for specific circumstances. These are given in the books.

Plates that are supported on opposite ends act like bars. If they are also supported on the sides, they have higher buckling strengths and deform into waves or quilts. There may be several waves along and across the plate; this is a function of the length-to-width ratio and the thickness. Again, there are equations giving the strengths in the literature. The stresses at buckling may be elastic and below the yield stress so that when the load is removed, the waves disappear, no damage being done.

Cylinders also buckle into waves. They are sometimes stiffened with rings, either inside or outside. This type can buckle between rings which themselves may stay circular or the rings may buckle too. This needs an expert's solution. Cylinders under external pressure loading may buckle catastrophically with some violence, the deformations being permanent.

4. Fatigue strength

Fatigue is probably the most common failure suffered by structures. In addition, it is often the most difficult to predict in a simple way. It is different from those considered so far because in metals the fracture starts from a very small fault and develops into a crack which grows over time, until an ultimate failure occurs as the material which resists load reduces. If the fracture face of the crack is examined after failure, curved striations can often be seen over part of the face. These trace the progress of the crack front and emanate from the original fault. The remainder of the face exhibits the usual tensile overload type of fracture. The source of the faults will be discussed later. Fatigue is predominantly a tensile stress phenomenon. The crack is pulled open a little every time the load is applied.

The first thing to understand is that the general stress level at fatigue failure is less than the yield or proof stress. The stress level at failure varies according to how often the load cycle is repeated. Thus the two important items of data required in each load case are the stress level and the number of occasions this stress is reached, known as the number of stress cycles. The higher the stress, the fewer cycles before failure. For steels, this is generally true down to a certain stress level known as the *limit strength*. Below this, the component can apparently endure an infinite number of cycles. Most other metals and their alloys cannot. The fatigue strength is always lower at a higher number of cycles. Consequently, for the other metals, it is usual to quote the fatigue strength at ten million, one hundred million, or five hundred million cycles (that is, at 10^7, 10^8, or 5×10^8 cycles). At this level, the change in strength will be very low as the cycles increase.

Obviously, cycles have to be counted or estimated some way. They may accumulate slowly or very quickly. For instance, turning the key of your car when you start it is done once per car use (assuming you open the door electronically). If the car is scrapped at twenty years old and is started five times a day, on average the key suffers $20 \; years \times 365 \; days \; per \; year \times 5 \; times \; per \; day = 36500$ cycles. As for the wheels of the car, if they complete two hundred thousand miles at one thousand revolutions per mile, they suffer two hundred million cycles, which is a lot more. The same fatigue allowable strength could not be used (unless the applied stress was very low), assuming both are made of the same material. If a structure is vibrating at many cycles per second, it is obvious that even more cycles would have to be catered for.

Each load case must be examined for a number of conditions appertaining to how and where and in what way it operates. In addition, the fatigue strength of any component is dependent on the material it is made of as well as its method of manufacture – that is, its final condition. These will be considered in turn.

Mean and alternating stresses

One of the most important things is the make-up of the loading. Mostly there are two parts to a cyclical load and the stress it causes. These are the *mean* stress, which does not vary, and the *alternating* stress, which obviously does. The reason this division is important is that the alternating stress does most of the damage. The mean stress is the average of the maximum and minimum stresses in the cycle, while the alternating stress is half the difference of the maximum and minimum stresses in the cycle.

The mean stress is

$$s_m = 1/2 \; (s_{max} + s_{min})$$

The alternating stress is

$$s_a = 1/2\ (s_{max} - s_{min})$$

The s_{max} and s_{min} are the maximum and minimum stresses in the cycle. The loading system is illustrated in the figure below.

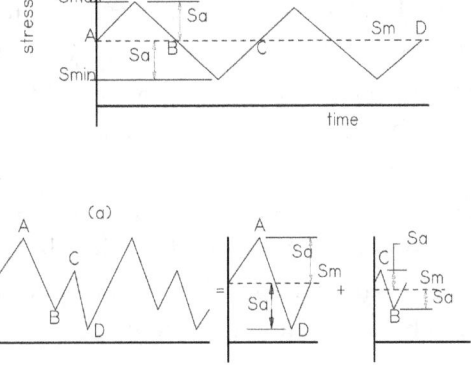

Figure 8.2 Typical cyclic stress cycles versus time

Figure 8.2 (a) shows a trace of stress in some component varying over time between s_{max} and s_{min}. Applying the above formulae gives s_m, shown by the dashed line, and s_a, which is the stress varying equally above and below the mean line. The trace is shown as varying linearly, but it could be curved or could change suddenly; it does not matter. The important aspect is the maximum and minimum stress levels. These define the cycle. A cycle starts at any stress and is complete when it next reaches that point, going in the same direction.

Figure 8.2 (b) shows a complication as the stress trace varies more. The technique here is to find the maximum and minimum stresses and use these to define the largest cycle, defined by the maximum, A, and minimum, D. There is then another smaller stress cycle – maximum, C, and minimum, B. There could be more cycles that are different. Each maximum and minimum could be made up of several load cases. It is important to find the largest figures. Both cycles do damage, but most is done by the larger if it occurs as often as the smaller. Therefore, that is the one to concentrate on initially, to find out whether a problem exists. It is possible to assess the combined effect of the two cycles, but it needs a specialist, for an analysis called Fracture Mechanics is needed, which will not be dealt with in this book. Simple methods such as Miners Rule, are not very reliable.

Some loading patterns are not regular; they exhibit randomly sized variations. Examples are car wheels going over bumps or aircraft flying through wind gusts. The loads can only be predicted in size and number by using statistical probability methods. This is the work of experts.

Now that we have the number of cycles with the largest alternating stress, we must find the allowable stress for that situation. If the material is steel, the first thing to do is to see whether the maximum applied stress is below the fatigue limit strength. If it is, all is well. If not, and the steel is ordinary carbon steel, it is worth knowing that it takes about one thousand cycles to the yield stress to fail a component. For other metals, if the stress is well below the failure strength at 10^7 or 10^8 cycles, then the situation is satisfactory. Otherwise, for all materials, a Goodman diagram will have to be constructed.

This is drawn with the mean stress plotted along the horizontal axis and the alternating stress up the vertical. See the figure below.

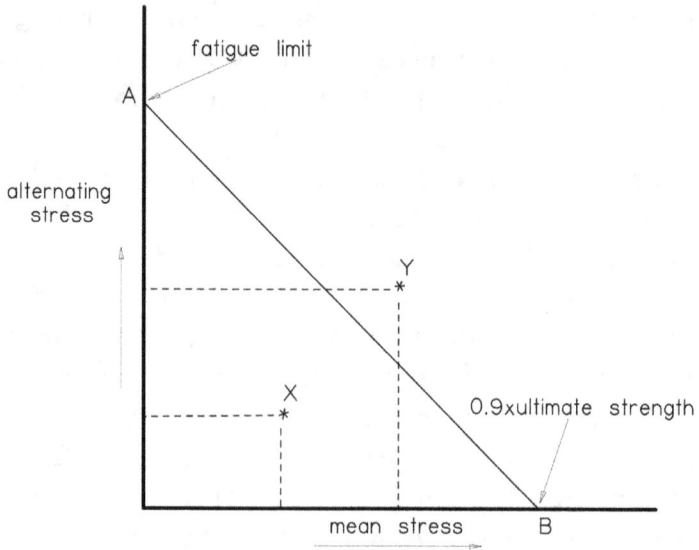

Figure 8.3 Typical Goodman diagram of mean against alternating stresses

Two points are known. The first is the ultimate strength at B, nowadays conservatively valued at 0.9 times the ultimate stress. Here the alternating stress is zero and there is no fatigue issue, so the failure strength is the usual static values. The other point is the fatigue limit strength at zero mean stress, point A. This is when the stress is fully reversed so that each point repeatedly reaches its maximum tension stress and then an equal compression stress. In order to find this fatigue strength, special tests have to be done. The basic tests are performed in benign conditions (see later) with zero mean stress. This can be achieved in a number of ways. A flat bar can be bent up and down equally, a bar can be pushed and pulled equally, or a special test piece can be rotated with a constant force applied, the so-called rotating bend test. The last is the usual; it's easier and cheaper.

Using a series of test pieces which are specially prepared, a graph is drawn which plots each resulting failure stress against the number of cycles to failure, a so-called S/N curve, as in figure 8.4.

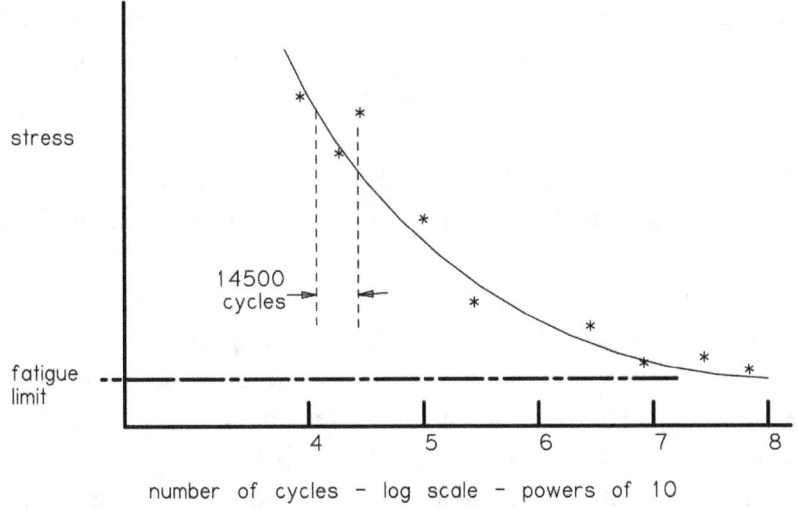

Figure 8.4 Typical S/N curve with a few test results

This diagram is a typical representation of a large number of tests on a particular material in specified conditions. It is not an actual result. Because of the numerous cycles, one hundred million, they are plotted on a log scale. This gives the same interval for each factor of ten. The curves are usually of this shape. For carbon steel, the curve levels out to a value which, of course, is the fatigue limit. Other materials have a continuously falling curve, which is why a number of cycles must be quoted if an allowable is to be specified. This value is quoted by material manufacturers. The fatigue limit is used to construct the Goodman diagram.

The asterisks (*) are individual results and are not exactly on the line, which is an average smoothed representation. This shows a characteristic of fatigue failures – i.e., in apparently identical circumstances, the failure stress varies quite widely. The scatter in results can be five; some specimens last five times longer than others. In the figure, the gap between the average curve and one test result is 14,500 cycles (between 11,200 cycles and 25,700). The scatter may be because it is impossible to get two tests performed identically; but then, in real life, two situations are never identical. A large reserve factor is required.

Returning now to the Goodman diagram, figure 8.3, a straight line, AB, can be drawn between the fatigue limit, alternating stress value, and 0.9 times the ultimate mean stress. Any component with stress values that plot below the line is safe, but values that plot above the line are not. X is safe; Y is not. The line AB may not be straight. If the material is ductile, it may be a parabola above the straight line, showing that this is more than safe. However, if the material tends to be brittle, even the straight line may not be safe. Specialist opinion should be sought.

As plotted, the mean stress is tension. If the mean stress is compression, the fatigue limit is better, as is implied by the line B to A extended leftward. However, it may not extrapolate linearly, so only the knowledge that the situation is safer can be inferred unless better information on the particular material is available.

Much information is available online. Google "typical fatigue curves metal" and you will find a Wikipedia entry, which is a good starting point. There are dozens of examples – one of these may be useful if you know the material type and grade of your structure.

SCF

If the component under consideration has a stress concentration, then the safe thing to do is to use the peak stress in the fatigue calculation. In practice, it has been found that the fatigue strength is higher than this prediction by a factor between 1.1 and over 3. This range has been collected in a book called *Metal Fatigue*, by Frost, Marsh, and Pook, a very detailed study of the subject. The factor depends on the material, the severity of the SCF and the type of stress (pure tension or bending). If the component is in trouble, expert advice should be sought.

Surface conditions

Since fatigue cracks mostly start from the surface (the exceptions being from internal flaws such as casting inclusions or voids), the condition of this is very important. Some are detrimental and some beneficial to the fatigue strength. The best superficial condition is a smooth surface since this inhibits crack initiation, so it follows that a rough surface is bad.

Coarse machining, casting, forging, and extruding are all manufacturing processes that produce poor surfaces. For best fatigue life, they should be ground and even polished. In use, surface damage by corrosion, fretting, or indeed any indentation damage will induce cracks and shorten life.

In contrast to these life-reducing effects, if the surface layer, perhaps only a few micrometres in depth, is in compression, obviously the tension due to loading is reduced and the life is enhanced. This tension may be induced as a by-product of the manufacturing procedures or deliberately induced by processes such as shot peening.

Stress corrosion

Worse than a corroded but dry surface is a component in a corroding or wet atmosphere. Here the corrosion and crack growth mechanism combine to reduce fatigue life more than in dry conditions. This condition needs oxygen and water, seawater, or a similarly corroding medium – for instance, any components at sea where the very atmosphere is damp or which is subject to spray.

In both of the last sections, the fatigue limit strength must be established by special tests which simulate the required conditions. Strength figures are published, many of them up to 1973 in the above-mentioned book. Specialist literature and manufacturers specifications should provide more, although it is sometimes difficult to pin these down to one's own needs. If in doubt when your gismo's stresses seem relatively high, seek help.

Multiaxial stress

If a component is loaded from several directions so that the stresses at a point of interest on a surface are in two directions at right angles to each other, the question arises as to how best to combine them before dividing them into mean and alternating parts. In part 5 of Chapter 7, there was an equation giving the principal stresses from the applied stresses. However, for failure cases, different criteria apply, and the equation that fits the failure data from tests best is due to von Mises; it is also known as the maximum shear-strain energy failure criteria.

$$\sigma_e = (\sigma_x^2 - \sigma_x\sigma_y + \sigma_y^2 + 3\tau_{xy}^2)^{\frac{1}{2}}$$

Herein, σ_e is the stress equivalent to the applied stresses, which are σ_x, σ_y, the orthogonal tension or compression stresses, and τ_{xy}, the shear stress. The σ_x is either the maximum or the minimum stress from the fatigue cycle (see Figure 8.2). The σ_y is the associated orthogonal stress and τ_{xy} the associated shear stress. All these stresses are surface stresses – that is to say σ_z, the stress into the component, be it bar or plate or whatever, is zero or negligible. This analysis is the usual one for the conditions at the onset of yield and is assumed to be valid at lesser stresses at the suggestion in *Metal Fatigue* by Frost, Marsh and Pook.

There will be two of these σ_e values, for the maximum σ_x with its σ_y and the minimum σ_x with its σ_y (or the other way around if σ_y is greater than σ_x), and these give the maximum and minimum of the equivalent stress cycle. They can be split into the mean and alternating components and a Goodman diagram constructed.

Temperature

When a structure is heated, there are two possible effects. The first has been dealt with; it is the stresses and strains which occur due to the expansion of the materials if the structure is restrained in any way. This amounts to a load case. The second is the effect on the strength properties of the material. Fatigue is affected. In general, most materials have a higher strength at lower temperatures than at manufacture (usually described as room temperature, or RT). They may have a somewhat better strength up to one hundred or two hundred degrees Celsius, but then the strength deteriorates with increasing temperatures.

Actual changes in strength are dependent on the precise make-up of the material, and specialist help will have to be sought. Indeed, tests which replicate the relevant conditions may have to be conducted. For some steel alloys, the improvement with an increase in temperature can be 100 Mpa, and with others the fatigue strength falls from RT, at first gently and then severely. Aluminium and magnesium alloys usually lose strength, as do plastic composites. Metallurgically stable metals can double or more their fatigue strength at low temperatures such as -200°C. Fatigue is a specialist subject which is nowadays dealt with by assessing and measuring crack growth under the heading of Fracture Mechanics.

5. Creep strength

This is the phenomenon of continuously increasing strain in a structure when it is under a constant load at some temperature. Unreinforced plastics creep at normal room temperature. Most metals creep at much higher temperatures. In plastics, the mechanism of creep is due to the long chain molecules which are entangled around each other, slowly sliding past one another. This is at least partly a permanent effect, although there is some recovery when the load is removed. In metals, there are several mechanisms. These are to do with the grains which make up the metal at the microscopic level. These may slip past each other or their internal structure may change.

In order to assess whether a metal component under load at high temperature is likely to creep, look up the melting point on the absolute temperature scale (add 273 degrees to the Celsius temperature to get the absolute) and divide into the load temperature, also on the absolute scale. If the answer is more than 0.5, you are in the regime. The trouble is that data on creep rates is scarce, or only known to the relevant expert. Also, your structure will eventually fail, so the total time at high temperature and load will have to be counted, an expensive nuisance. This is the work of experts.

Stress relaxation is the reverse of creep. By some mechanism, the structure, under high temperature and load, is kept from deforming; the stress in it reduces with time, rapidly at first and then more slowly and asymptotically to a low value. This is important in bolted joints where the bolt is stretched elastically to provide the friction between the threads necessary to stop the nut from coming loose. Other mechanisms are sometimes held together by interference fits, so a pin may be forced into a slightly undersized hole; these are obviously also vulnerable.

6. Stress corrosion cracking strength

If a component is in unchanging, or static, tension and in a particular corrosive environment, it will fail at a lower stress than the yield stress. For metals, if the material is an alloy, each alloying element is vulnerable to certain corrosive substances – for instance, an aluminium alloy and seawater. There are many combinations. Non-metallic materials suffer similar failures.

The damaging corrosion occurs at the internal tip of the corrosion crack, where the stress concentration is large. This produces conditions conducive to further corrosion, and new bare metal which can corrode and therefore a larger crack, and so on.

There is a stress level below which the component will not fail. However, this is not the whole story, and the problem is best described by a subject called Fracture Mechanics which is outside the scope of this book. (The actual critical stress is described by the so-called stress intensity). Again, it is necessary to consult specialists.

7. Limiting Deflections

In some structures, it is necessary to limit the deflections under load. This will be for functional reasons. The Building Regulations, for instance, limit the sagging in a floor to 0.3 per cent of the span not for strength reasons but because the floor would be disconcertingly flexible and bouncy; people would not feel safe even though they were. In machinery, care must be taken so the load-induced deflections do not lead to clashes of moving parts and so forth.

Very often when a structure is constrained by this consideration it is found that the stresses in it are low. It is a more severe case. Clearly each and every structure has its own deflection limitations which will have to be discovered individually. As usual, Safety Factors must be applied.

8. Vibrations

Some simple equations were given in Chapter 7, section (e), so that a preliminary idea of the fundamental natural frequency could be gained. The question arises: what is to be done with this information? It is needed if there is a load on the structure which is repetitive. If this load repeats in the same time interval as the natural frequency of the structure, then the deflections of the structure will increase, theoretically without limit. Actually, some form of damping will limit the vibration amplitude (or size) if something does not break or severely deform some part of the structure first. This damping will also cause the vibrations to die away when the loading is removed.

Even if the periods of the natural frequency and the forcing load are not exactly the same, although near, the deflections of the structure are still increased above what they would be under static conditions. It is a good idea to have a minimum margin of 30 per cent between the two frequencies. Obviously, it is necessary to have knowledge of the value of the forcing frequency.

A single impact on a structure may also cause vibrations at the natural frequency, which may be damaging. Sometimes, of course, such an impact is deliberate and to the purpose, as in a drum.

A number of changes to the structure can be made in order to avoid the coincidence of the forcing and natural frequencies. The mathematics describing vibrations in structures shows that the natural frequency is proportional to the square root of the stiffness of the vibrating part and inversely proportional to the square root of the mass. In other words, if the stiffness is increased by 2, then the natural frequency is increased by root 2, which is 1.41. If the mass is increased by 2, the natural frequency is *decreased* by 1.41. The stiffness of a beam is a function of the second moment of area of its cross section, the length, and the material; of a plate the thickness, dimensions, and material. Also, the constraint conditions will influence the result.

If this is not possible, then dampers can be added, as in the Millennium Bridge in London. This involves even more mathematics and is best left to experts.

9. Common materials' strength

Listed in the table below are some typical strengths for various materials. These are a very few out of literally thousands. They illustrate the level of allowable strengths for each type of material you might expect, but it must be emphasised that they do not necessarily apply to your project. For this reason the material specification is left vague. I am afraid you will have to discover precise material specifications before assigning more accurate strength figures to them. All the figures are conservative.

Material	Ultimate tensile strength	Yield or 0.2% proof strength	Fatigue limit as ratio of UTS*	Young's modulus	Notes
	MPa	MPa		GPa	
Metals					
Steel – mild	465	230	50%	210	1
- high tensile	1700	–	41%	210	2
Cast iron	310	–	42%		3
Aluminium – pure	150	140	65%	70	
Al/copper alloy	485	425	60%	70	
Al/zinc alloy	570	505	60%	70	
Al/magnesium and silicon alloy	310	260	60%	70	
Composites					
Polyester and 30% w glass – random lay-up	85	–	70%	7	4
Epoxy and 70% w glass – uni-dir. lay-up	930	–	33%	29	4,5
Epoxy and 50% carbon fabric	350	–	50%	85	4
Epoxy and 60% carbon uni-dir. lay-up	1000	–	33%	175	4,5
Unreinforced plastic					
Nylon 6	70/85	–	20%	1.4	6
Polypropylene	21/37		20%	1.3	6

Notes

1. This is a typical low strength steel in general use.
2. This is a much higher strength material used for springs and other special purposes. The figures are taken from the book on fatigue by Frost, Marsh, and Pook. Proof strength is not specified but will be about 90 per cent of ultimate.
3. The proof stress of cast iron will be very near the ultimate and is irrelevant for stressing purposes.
4. These composite examples give an idea of the range of figures. You can design for almost anything in between.
5. The values are for the most favourable direction. These fall severely as the direction changes.
6. These are just samples of the thousands of plastics known to man nowadays. Even these specific types of plastic are subject to manufacturing processes – you may get less from some sources. Also, temperature and relative humidity change the values significantly. The numbers come from *Plastics Engineering, Manufacturing and Data Handbook* from the Plastics Institute of America.

* Except for steels, the figures are for 10^7 or 10^8 cycles to fatigue failure.

10. Bolt allowable stresses

Except in special circumstances bolts come in standard strengths. The strength (both tensile and shear) is obviously a product of the allowable stress and the cross-sectional area. Both of these are of commonly accepted values; they are not of any old diameter but in set ones, usually from 1.6 mm to 68 mm in twenty-nine steps. The material has specific values from 386 N/mm² to 1200 N/mm² in nine steps. This gives a large variety of bolts, of which some are preferred and therefore more easily obtainable and cheaper. Because of this standardisation, you should be able to go out and buy reliable bolts from any manufacturer.

Since you cannot necessarily tell by looking at it what material a bolt is made from, they are mostly marked. The British marking system has changed to bring us into line with the EU, so there are two systems still around. The new system is used in all new design, but if you are studying an old – say, 1980s or before – structure, you may come across the old markings.

Non-corrosion resistant bolts or screws have a designation which looks like this: n.m, where n is a number giving 1/100 of the ultimate strength of the bolt material in N/mm^2 and m is the ratio of the yield strength of the material to the ultimate. For instance, 6.8 indicates an ultimate strength of $600N/mm^2$ and a yield strength of $0.8 \times 600 = 480N/mm^2$. The figures vary from 4.6 to 12.9 in 9 steps. These are the standard bolt strengths you can obtain. The ratio of 0.9 implies that the material is nearer the brittle end of the spectrum. This number normally is stamped on the bolt head. There will also usually be a manufacturer's mark.

If you cannot see the bolt head in an existing structure but you have a portable hardness tester that you can use on an exposed part of the bolt, you can estimate the bolt type from the hardness. This is a bit approximate because the hardness values overlap the bolt strengths.

Table 1

Vickers Hardness	120–220	130–220	155–220	160–220	190–250	250–320	290–360	320–380	385–435
Property class	4.6	4.8	5.6	5.8	6.8	8.8	9.8	10.9	12.9

Vickers hardness versus bolt property class

You can see that a hardness reading of 210 means you should choose the $400N/mm^2$ bolt strength, even though it may be a $500N/mm^2$ bolt – it's similar with the higher values. There are several methods and scales of hardness testing and there are equivalence tables in such literature as *Machinery's Handbook* and the British Iron and Steel Producers Association specification handbook.

On small bolts, a system based on a clock face is used.

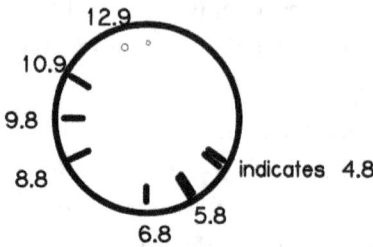

Figure 8.5 Markings on small carbon steel bolts

In the case of 4.8 and 5.8 bolts, there may be only one dash. This indicates that the bolt material is 4.6 and 5.6, respectively. The dot shown at the top is to indicate twelve o'clock on the clock face. The dot to its left is to indicate 12.9 strength material. The twelve o'clock reference point could also be a manufacturer's mark.

Studs have their own markings, shown in the table. These should be on the exposed ends. They may be on the shank, and as they would be indented, they may cause a stress concentration which is a bad idea in fatigue cases.

Table 2

Property class	5.6	8.8	9.8	10.9	12.9
Symbol	—	○	+	□	△

For stainless steels, there is a different system. The types of steel are split into three, depending on their metallurgical make-up, then each into several more depending on their alloying elements; and there are four strength levels. The types are austenitic, martensitic, and ferritic, which does not concern the stressman, as the designer will choose it for other than strength reasons. The exception may be austenitic, as at the low strength level, it is non-magnetic and so will be found on electronic devices.

The marking is of the form Np-m. N is either A, standing for austenitic, C for martensitic, or F for ferritic. The p is a number 1-5 (describes chemical composition) and the m is 50, 70, 80, or 110. This last is the important information, for it is one-tenth of the ultimate strength of the material. So an A2-50 is an austenitic stainless steel bolt of ultimate strength 500 N/mm², while a C3-80 is a martensitic not so stainless bolt of strength 800 N/mm². The following table gives the list of bolts, their strengths, and the hardness number (more on that later).

Table 3

Type	mark	Ult. N/mm²	Yield N/mm²	HB	HRC	HV
austenitic	A1-50, A2-50	500	210			
	A3-70, A4-70	700	450			
	A5-80	800	600			
martensitic	C1-50	500	250	147/209		155/220
	C1-70	700	410	209/314	20/34	220/330
	C1-110	1100	820		36/45	350/440
	C3-80	800	640	228/323/209	21/35	240/340
	C4-50	500	450	147/209		155/220
	C4-70	700	410	209/314	20/34	220/330
ferritic	F1-45	450	250	128/209		135/220
	F1-60	600	410	171/271		180/285

The old marking systems, described below, should cover most other eventualities you may encounter in existing older structures. There are two issues: the strength of the material and the cross-sectional area of the bolt at the bottom of the thread. The strength designation is in the form of letters A, B, P, R or S, T, V, X. Each defines a strength grade and is stamped on the bolt head.

Table 4

Grade	Ultimate strength		Yield strength		HB
	Tons/in²	N/mm²	Tons/in²	N/mm²	
A	28	432	–	215	126
B	28	432	–	215	126
P	35	540	–	420	152/240
R	45	695	34	525	201/285

S	50	772	38	587	152/240
T	55	849	41	633	248/335
V	65	1004	52	803	243/370
X	75	1158	63	973	341/410

There are also Black bolts which have no markings but have 25 Tons/in² or 386 N/mm² ultimate strength and 200 N/mm² yield. In fact, for safety's sake, any unmarked bolt should be taken to have only these strength values. The yield figures in italics are suggested safe values taken from general literature.

The second consideration is the cross-sectional area of the bolt, and this depends on the type of bolt. Over the years, there have been various standards to which the threads have been formed; the first proper manufacturing standard was BSW (British Standard Whitworth). BSF (British Standard Fine), while UNC (Unified Coarse), UNF (Unified Fine), and UNJ are the most common, but there are others, such as pipe threads, square threads, and so forth. The crucial point is that for any given nominal bolt size, the core diameter varies with these differing standards. Therefore, the cross-sectional area varies and so does the strength. The quickest, if conservative, way to calculate the strength is to take the smallest minor diameter of the thread of the various standards for a given nominal bolt size. On this count, the old BSW has the smallest minor diameter. Nowadays it is easiest to Google details of these threads, but there are complications, as the pitch of the threads or the number of threads per inch may also have to be known. The upshot for bolts of unknown provenance is to use Whitworth thread details and twenty-five tons/inch squared strength; you will be safe. A note of caution: some bolts have shanks between the thread and head which have been turned down to a small diameter for a design reason.

Where the bolt is in shear, you need to know the allowable shear stress. This is not often quoted in the literature, as the situation is complicated. The circumstances of the loading have to be considered. A safe value for the yield shear stress is 57 per cent of the tensile yield stress.

The source of the above information is various British Standards contained in a very useful BSI handbook – No. 44 *Threaded Fasteners.*

US data can be obtained from *Machinery's Handbook* (and Google, of course).

The hardness numbers might be useful for identifying the strength if all else fails. The range given in the grade table above covers most steels. You will need a hardness tester, and portable versions are available (Google them). Doubtless there are also consulting companies that will do this for a fee. Note that the hardness numbers are ranges, as the correlation between strength and hardness is not exact. Aluminium and its alloys have different scales, but the correlation between hardness and strength is more difficult; a specialist is required.

Markings are either embossed or indented.

Other countries have their own marking systems.

Section (c) Reserve Factor

Having done all this work, you need to summarize the results and get some overall idea of how strong your structure is. The Reserve Factor, or RF, system gives this summary and records where the weakest points in the various components of your structure are situated. By consulting this document at future times, anyone can see what to do if, for instance, load cases change.

This is an ideal situation which is only achieved in well-regulated industries – for instance, the aircraft industry, the nuclear one, and so forth. Mostly designers do not even seem to keep their calculations. However, it is better to be disciplined, and it is easy to do because it is just a matter of keeping a list of work done and results found. Make a table with perhaps a dozen headings and list each critical or high stress in each component that carries a reasonable load. If the stresses are very low, they can be ignored; many components can also be neglected as being low-stressed.

Different projects may require different treatment. The first thing that is important is to identify the component and the point on it where the stress in question is located. Table headings might be "component," "drawing number" (if you have one), "position on component". It may be advantageous to produce a sketch in order to clarify the exact point of the high stress.

Next, the material should be specified as closely as possible. Just steel or aluminium or whatever is not very helpful, although it is a start and may be all that is known. But remember that some steels are ten times stronger than others, and the same goes for other materials. If the exact specification is not known, conservative assumptions must be made that the steel is of minimum strength, say 240 N/mm² at yield and 400 N/mm² at ultimate. If this is not enough to pass the component as strong enough, it may be possible to check the hardness of the material if it is a metal. There are portable hardness testers and information on the correlations between strength and hardness available.

The load case must be specified, probably by reference to another list that has been made of all of them. Then the type of stress, bending, shear or whatever, and whether ultimate or fatigue should be noted.

Now we get to the nitty-gritty, the actual stress you've calculated and the relevant allowable stress. The Reserve Factor, or RF, is the allowable divided by the actual.

$$RF = \frac{allowable\ stress}{actual\ stress}$$

Instead of stresses, loads could be used. For instance, in the case of bolts, this would be usual.

If the piece is strong enough, the RF will be greater than 1.0. If not, it will be less, of course. However, in the happy case of the former, the actual value will tell you how much you have in reserve for overloads. If you have a complete listing for a complicated structure, you can run your eye down this column and quickly see which point is critical, which is a useful thing to be able to do. Americans use a slightly different version of the above, called a Margin of Safety or MS. This is the RF minus 1.0, so a positive MS means the component is strong enough but a negative one shows it is inadequate.

Finally, it is worth noting where your calculations are kept, including any Finite Element models and results you might have had done for you. This is especially true if you have many papers. The very last column should be for comments. Here you can note anything about the calculation (conservative or not, perhaps) or anything else relevant.

CHAPTER 9

Examples

Finally, we will work through a few examples, as they might be thoroughly and formally done, in order to bring the various parts of the book together. I have chosen some everyday items, not things that look industrial, so that everyone can relate to them more easily. This choice should also show that anything can be stressed; whether it should be depends on the consequences of structural failure and on how different it is from established similar items.

The following calculations are all simply the choosing and the applications of simple equations which describe the problem. All that is required mathematically is the substitution of numerical values into the symbols in the equations and then calculating the result – simple. Choosing which equation and exactly what its result means requires more thought. Hopefully the examples will illuminate the processes required.

I have considered the impact of each chapter consecutively; this avoids forgetting important matters. It is in the order in which any analysis ought to be carried out.

1. A chair

Ch 1   Load cases

So how does one generate the load cases when the gismo is new and/or there is no help from codes of practice? To take a simple example, think about a chair, any sort of chair. Someone has written a specification which is mostly about the size and type, whether it is a dining chair, an armchair, a recliner, or whatever, in addition to the furnishing, colour, and other irrelevant information. He might also have specified the heaviest person to sit on it. This piece of information is what we want. If such a document does not oblige, then a figure will have to be estimated. Anyway, we have a load case! It is due to mass, our first load type. It may be enlarged upon by considering how the person sits; will he ride back on two legs if this is possible and then lean over to one side? If he can, someday he will, and this means that most of his weight is on one chair leg. This might also come under the heading of foreseeable misuse, our last load type.

Notice that a failure may not be too serious in this example. If it were, by causing lasting injury or even death, then a more rigorous attitude should be taken; an effort must therefore be made to discover the heaviest person ever to sit on the chair, probably by consulting medical evidence. It may be an academic point whether the result of a broken chair leg is serious or not. It will depend on the chair construction, but it may be safer to assume that it is always serious.

Another load case concerning armchairs is that of someone sitting on the arm. Special chairs, such as recliners, will have to have a more detailed analysis of the case. The load in each part of the reclining mechanism will have to be found. The second and third load types, rotation and fluid pressure, seem unlikely to be relevant.

Nor is there is likely to be a load case due to temperature, unless it is a plastic chair in a very hot climate and the plastic cannot stand the temperature, but that is a wrong choice of plastic. Note, though, that unreinforced plastics lose strength with increasing temperature and that this happens at values not much higher than room temperature.

There may be a vibration case. Either the occupant jiggles a lot – say, a small child – or the chair might be on a vehicle which is vibrating.

A moving load case is not applicable, but impact or shock loading should be thought about. Chairs, except office ones, are usually stationary, so they should not have crashes but some people drop into a chair. This increases the effect of their weight by quite a lot.

We have thought of the misuse case of riding back. Is there a failure case? Again, depending on the construction, it may be reasonable to define one. If it is a four-legged dining chair, it might be stipulated that the remaining legs should not fail also because of one breaking. If one link in a recliner's mechanism fractures, there should not be a catastrophic collapse of the chair. The rest of the mechanism must support the sitter even if it jams.

Ch 2   Safety Factors

For chairs in all general use in house, office, or outside, there is no point in choosing low Safety Factors, even if the design is to be elegantly slender, because in structures in the process of being designed, low RFs will impose restrictions on subsequently necessary or desirable changes. Anyway, a little extra strength to reinforce critical areas is well worth the extra weight; also, there will be fewer restrictions on the method of manufacture – a cheaper, less well-controlled procedure can be accommodated. So for the yield and ultimate cases in metallic structures, assuming that safety is involved, use at least 2 and 3, respectively. For any metalwork that is cast, choose also the casting factor of 1.6. This gives total factors of 3.2 for yield and 4.8 for ultimate. In the case that fatigue occurs in a component, keep the stress below the fatigue limit and put a factor of 2 on that. If unreinforced plastic is used, the factor is 10. If you have the opportunity, it is also worth designing the components so that there is no or minimal long-term stress in them. This avoids the problem of relaxation (see later).

There are seats which are very carefully designed in reinforced plastics – for instance, aircraft seats – because they have arduous duties but need to be lightweight. They might use an ultimate factor of 1.5. This is outside the scope of this book.

Ch 3   Structural Data

The size of an existing chair and its components, seat, legs and any supporting struts, back, and arms can easily be measured or taken from manufacturer's drawings. At intersections, measure the length of struts and legs from the middle of the joint.

The material type of an existing item is more difficult to classify. It should be easy to decide between steel, aluminium, wood, plastic and ceramic, but exactly which grade cannot be established by sight. Some hints as to what to do are given in Chapter 3.

Some representative strength values of various materials are given at the end of Chapter 8, section (b). These are the lowest in each category. While using them, if you find that your structure is not strong enough theoretically, you may find a stronger grade of material that is good enough if you are making the chair from new. You will have to establish the exact grade of an existing component somehow if the lowest is not good enough.

Ch 4   Classification by component type

Whatever the type of chair, it will have some legs and surfaces to support the person. The seat can be analysed as a plate with supporting beams on at least two edges and the legs as struts and/or beams. In the case of an office swivelling chair, the vertical central support is a strut and the five horizontals are beams. The back will bend, so it is also a beam.

Ch 5   Support Reactions

The reactions to the weight of a chair and anyone who sits on it obviously apply to the chair's feet. It may be that most of this load is travels through just one foot. The exact detailed load path depends on the geometry of the chair. If the feet are symmetrical, as in the example shown in Figure 5.10, then that method can be used to find the distribution of the reactions at the feet.

Ch 6   Loads and the manner of their application

The seat of a chair will be loaded by the person's behind, which would be characterised as a pressure. The seat must be checked for the heaviest sitter, say, 150 kg (about twenty-four stone). This must be converted to Newtons by multiplying by the gravity acceleration of 9.81 (call it 10 if you are making a quick mental estimate). It would be fair to assume a uniform distribution of load over the area of the seat. If the seat measures 430 mm by 330 mm the pressure, call it p, is

$$P = \frac{150 \times 9.81}{430 \times 330} = 0.01037 \; N/mm^2$$

The legs may be splayed so that there is a component of the load at right angles to it. Ignoring friction between leg and floor (perhaps the polisher has just been round or the leg is supported on a wheel) this will bend the leg. This load will be a component of the leg load. If the angle between the leg and the vertical is A degrees and the load is L, then the horizontal component is

$$LsinA \; N$$

This causes a bending moment up the leg, which reaches a maximum at the first support. If x is the distance to this support, the BM is

$$xLsinA \; N$$

The arms and front edge of the chair are stressed as beams and the seat pan as a membrane if it is thin and so would take little bending; otherwise, it is a plate.

Ch 7   Calculations

When doing strength calculations, it is necessary to have foreknowledge of the allowable stresses in the material because when you have found an actual stress in a component, you need to know whether it is acceptable. These allowables are described in Chapter 8 so both Chapter 7 and Chapter 8 have to be read and understood together before actual calculations start. Bearing this in mind, we can now do some stressing.

As a specific example, let us take a typical garden chair made out of a one-piece plastic moulding. There is nothing special about this chair; it's just an ordinary looking four-legged one with a back and two arms.

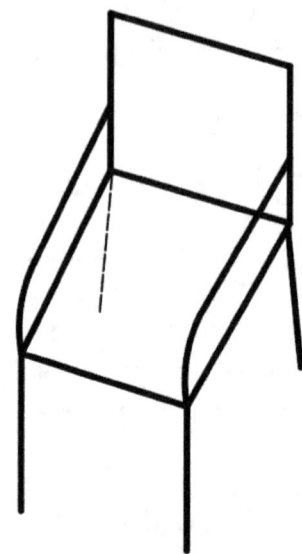

Figure 9.1.1 Line drawing of a simple chair. The lines depict the centroids of the components.

It is doubtful that any one has ever stress-checked a simple chair. I expect tradition has been followed – people just looked at what has gone before. If ever anything broke in a test, the size of the broken member was increased. We won't let that stop us; we will use the more modern scientific method.

It is useful to lay out the calculations in some ordered way, so first list the load cases.

Case 1.

This is a heavy person sitting quietly in the chair, upright, in such a manner that his weight is centrally placed on the seat. The calculation we have just done for a 150 kg person gives a uniform pressure of 0.01037 N/mm². Subsections of this case is when the sitter rides back on two legs and even leans to one side, perhaps to pick up something.

Case 2.

This arises from Case 1, when someone does not lower himself gently, which seems sensible in the case of a very heavy person who probably does not do much exercise and so will be thankful to relax into a chair. Assume that as soon as the person touches the chair, he relaxes and drops thereby doubling the pressure to 0.02074 N/mm².

Case 3.

This is the sitting on an arm case. Heavy persons probably would not try to balance themselves on the arm, as they are too ponderous and wary of doing such things. A teenager probably would do such a thing, so take a load of 75 kg. Using the equation in Chapter 6:

$$u = \frac{75 \times 9.81}{330} = 2.230 \; N/mm$$

The u is the loading along the arm, and the arm is assumed 330 mm long. Notice this loading is in N/mm, which is the total load divided by the length of the beam. In other words, we have a load of 2.230 N on every millimetre of the length of the beam. The total load is therefore the load/mm times the length of the beam. To digress and emphasise, we can have a load in N, a loading in N/mm, and a pressure in N/mm². It is important to understand the difference and bear it in mind when doing calculations.

Case 4.

This is the case of a heavy man standing on the chair. Such a situation loads the seat plate differently from sitting. He is reaching up for something nearly out of reach and so is on one foot, indeed on tiptoe, practically a point load, P. If we assume he is standing on the middle of the front edge, its BM will be largest. If he is nearly over one leg that will take most of his weight, but the BM will be low. For the former, the leg load is

$$P = \frac{150}{2} \times 9.81 = 736 \; N$$

Case 5.

This is the jiggling child case. The load will be much smaller, spread over a smaller area. We should look at it, though, as the allowable fatigue stress will be lower and the Safety Factor different. The situation is not immediately clear. For this example, take a child weighing 25 kg spread over a square area of 200 mm by 200 mm.

$$p = \frac{25 \times 9.81}{200^2} = 0.006131 \; N/mm^2$$

Note that the pressure is nearly as high as the heavy man's, but of course, the area loaded is smaller. For fatigue cases, the number of cycles of applied stress must also be calculated. If you have a fidgety child and he sits in this chair for a total of an hour for one hundred days over the summer and he moves fast, at the rate of five times a second, then there are

$$100 days \times 1\frac{hr}{day} \times \frac{3600 sec}{hr} \times \frac{5 cycles}{sec} = 1.8 \times 10^6 \; cycles \; that \; year$$

Notice how the units all "cancel" except *cycles*, which is what is required. This calculation tells us what allowable stress to look for on the cycles-to-failure or S-N curve.

Case 6.

For a failure case, let us consider a failure of a leg. The others should not break. In this case, if one leg breaks, the load will be distributed to two of the other legs, diagonally opposite each other, in an unequal fashion. We shall take a 75 to 25 per cent distribution, which seems reasonable.

Next we look at Safety Factors. It is to be assumed that garden chairs in unreinforced plastic come under the category of general engineering; there will be no official agency regulating their manufacture. The Safety Factor for static cases and unreinforced plastic is five for non-safety-related cases and ten for safety-related ones. Since falling off a chair due to a structural failure could be construed as safety related, we will use ten.

In the case of fatigue, we will try to keep the stress below the fatigue limit at 10^7 cycles and have a safety factor on the low side of the suggested range since there will be no safety implications – say, five, to be applied to the stress.

Now we have to find the structural data, but first the size of everything.

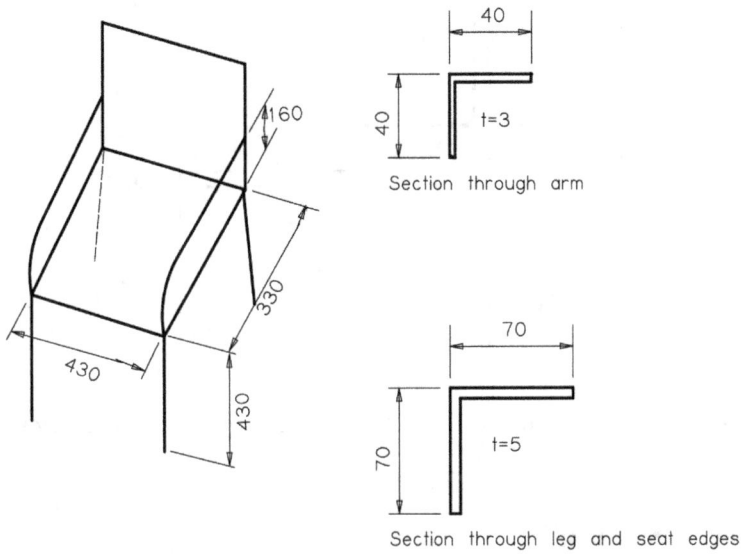

Figure 9.1.2 Dimensions of the garden chair, in mm

Assume that the chair is moulded from a polypropylene (a likely choice and its strength figures are given as examples in the tables) with an ultimate allowable of 37 N/mm² and a fatigue limit of 20 per cent of that (7.4 N/mm²).

Next the components to be stressed are the seat as a plate, the seat edges and the arms as beams, the legs as struts and beams. The required quantities are the second moments of area, or *Is*, of the two sections, the thickness of the seat plate, and the dimensions of the seat. If deflections are required, then the Young's Modulus will also be needed.

Referring to the addendum in Chapter 7(a), section 15, we can calculate the I required for each section shown in Figure 9.1.2. The arm is shown below, the leg and seat edge are similar except for the dimensions.

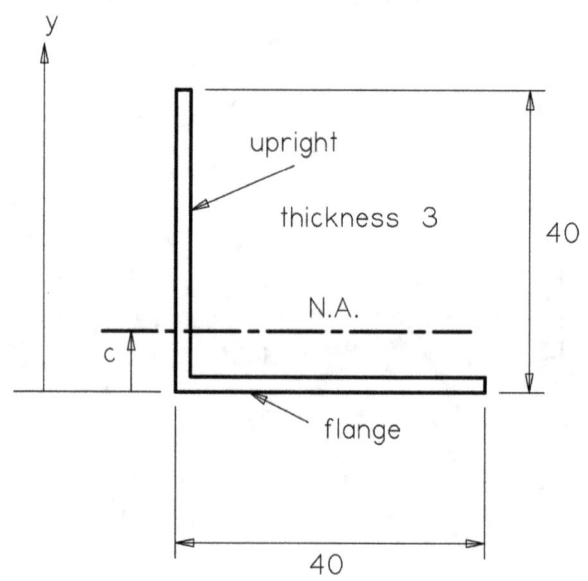

9.1.3 Section through arm labelled for the properties' calculation.

Section through arm

Item	Width, b	Height, d	y	A=bd	Ay	Ay²	I_s=bd³/12
upright	3	40	20	120	2400	48000	16000
flange	37	3	1.5	111	167	250	–
Sum, Σ				231	2567	48250	16000

Position of the neutral axis:

$$c = \frac{\Sigma Ay}{\Sigma A} = \frac{2567}{231} = 11.11$$

For the parallel axis theorem:

$$\Sigma Ac^2 = 231 \times 11.11^2 = 28513$$

The second moment of area we want is given by the following:

$$I = \Sigma Ay^2 + I_s - \Sigma Ac^2 = 48250 + 16000 - 28513 = 35737 \text{ mm}^4$$

Section through leg and seat edge

Item	Width, b	Height, d	y	A=bd	Ay	Ay²	I_s=bd³/12
upright	5	70	35	350	12250	428750	142917
flange	65	5	2.5	325	813	2031	–
Sum, Σ				675	13063	430781	142917

$$Neutral\ axis\ position, c = \frac{13063}{675} = 19.35mm$$

This is from the flange. From the edge, it is

$$70 - 19.35 = 50.65$$

$$shifting\ to\ the\ NA, \Sigma Ac^2 = 675 \times 19.35^2 = 252735$$

$$2nd\ moment\ of\ area, I = 430781 + 142917 - 252735 = 320963mm^4$$

It is worth looking at the numbers you calculate closely to see whether they are reasonable. In this case, it is possible to compare them; immediately we see that the second I is ten times bigger than the than the first. Can this be possible? Well, the thickness of the seat edge is 5 mm and the arm only 3 mm. From examination of the algebraic expressions, it is clear that I is proportional to t i.e. (5/3), which is 1.67; more importantly, I is proportional to the height, d, cubed (both y and c being functions of d) or (70/40)³, which comes to 5.36. The product of these is 8.95, near enough the ratio of the Is. This sort of quick calculation is comforting, suggesting that no arithmetical mistakes have been made. People experienced in particular circumstances (that is, who do similar calculations often) will get a feel for the numbers and know immediately if they have slipped up; the number will be wildly different from the expected one.

The next job is to find the support reactions for each load case. It is first necessary to consider the floor which supports the chair. Is it flat and firm? If so and the load is centrally placed, it can be assumed the load is split equally between the legs. This is an ideal situation, as has been said. However, the chair is constructed of unreinforced plastic, which means it is flexible relative to most floors. So it will flex to take up any unevenness in the floor. Of course, this will induce some stress in the chair. If we need to consider this – that is, if the chair is only marginally strong enough – the situation constitutes another load case. In this, we would have to find out what the stress was for a given – say, 1 mm – deflection of one corner of the seat. We shall see.

Load case 1

Sitting centrally, the 150 kg divides into four equal loads in each leg. At 9.81 N per kilogram, the load is

$$Leg\ load = \frac{150 \times 9.81}{4} = 368N$$

The subsections of this case (the person leaning so that most of the load is supported on one leg) would give a maximum load four times that:

$$Leg\ load = 1472N$$

Load case 2

In the case that someone drops onto the seat, we can double the load of the basic case 1. It seems unlikely that the entire load is carried by one leg alone, but it might be by two if the man sat on the forwards edge. The maximum leg load is therefore the same, that is twice the load but also twice the number of legs.

Load case 3

Sitting on the arm, obviously all the load is on one side of the chair and possibly more on one leg than the other, say, a 60/40 distribution:

$$Leg\ load = 0.60 \times 75 \times 9.81 = 441N$$

While this is small, it may have to be added to case 1, possibly twice the basic load since the sitting occupant may be leaning to that side:

$$Leg\ load = 441 + 2 \times 368 = 1177$$

Load case 4

The man-standing-on-the-front-edge load is 736N (as stated before) on each leg when he is in the middle of the edge, and this is not the highest. The reactions to each leg will be assumed to vary from a quarter of his weight to three quarters of it, so 368 N to 1101 N. More important will be the stresses in the front edge of the chair.

Load case 5

The reactions in this case are small since the wriggling child's weight is only 25 kg, which will probably be placed fairly centrally.

$$Leg\ load = 25 \times \frac{9.81}{4} = 61N$$

Load case 6

The broken leg case; we do not get the worst reaction from this.

$$Leg\ load = 150 \times 9.81 \times 0.75 = 1104N$$

We are taking the 75% of the load on one leg.

We need a summary so that we can look at all the results together. The reaction loads can, of course, be used as the basic loads to stress the legs.

Support reactions summary

Case	Loading	Load N	SF	Factored load N
1	Basic: heavy man leaning	1472	10	14720
2	Dropping onto chair	1472	10	14720
3	Sitting on arm	1177	10	11770
4	Standing on chair	1101	10	11010
5	Jiggling child	61	5	305
6	One leg broken	1104	10	11040

The biggest reaction is in the first two cases.

The first item to check is the leg. Case 1 says that the person is leaning such that all the weight is supported by one leg. The stress will be a direct stress, as any bending in the leg would cause him to topple over.

The stress is load divided by area:

$$14720/675 = 22 \ N/mm^2$$

The allowable is 37 *N/mm²*, so that is fine.

Case 2 is when the sitter drops onto the chair and the two front legs take the load. Now we have the case of a frame made up of the front edge of the chair and the two front legs. We can assume that the bottom of the legs are pinned by the surface on which it is resting; this will be grass or carpet, into which the foot sinks so as to trap it, or a harder surface which provides sufficient frictional resistance.

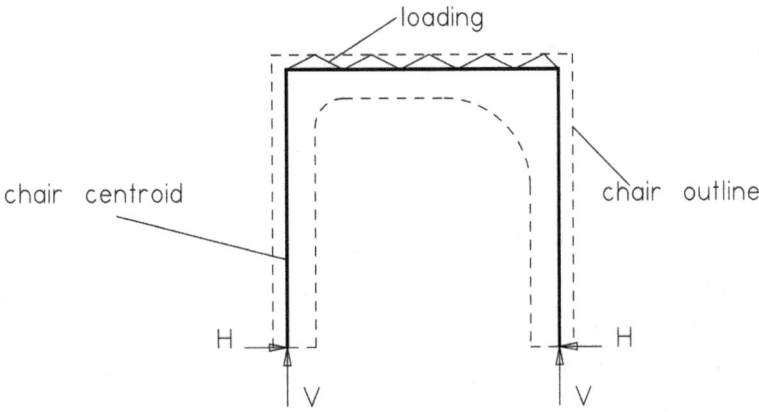

Figure 9.1.4 Front two legs and front edge of chair with UDL. The left-hand leg radius is smaller than the right-hand one as modified – see text.

In the drawing above, the centroid lines are as in Figure 9.1.2. The chair outline is useful because it shows where the leg cross section, as calculated above, actually ends, which is at the bottom of the radius of the inside edge. This is where the bending stress should be calculated. The loading is the 14720 N. V and H can be found from the usual tables (for example, *Roark*). V is obviously half the total. Since the leg and seat edge have the same length and sectional properties, the formula in *Roark* reduces to a simple one. The formula is long and complicated and involves ratios of the Is and multiples of the member lengths such that you can have three different Is and three different lengths. These all cancel if the Is and lengths are the same. The reduced formula is

$$H = \frac{W}{20} = 736\,N$$

W is the total load. The stressing then is:

$$Direct\ stress = \frac{14720}{2 \times 675} = 11N/mm^2$$

Remember that there are two legs and the area of the leg cross section is 675mm². This is a compression stress.

$$Bending\ stress = \frac{My}{I}$$

M is the horizontal reaction, H, times the distance up the leg. This is 430 mm to the seat edge centroid less 50.65 mm (seat edge centroid to the edge of the section – see Figure 9.1.3 and section properties through leg and seat edge) and also less the corner radius, which we will say is 50 mm to start with.

$$distance\ up\ chair\ leg = 430 - 50.65 - 50 = 329.35mm$$

$$M = 736 \times 329.35 = 242402Nmm$$

y is c calculated for the leg cross section earlier (19.35), that is the overall height minus c. The result is either 19.35 or 50.65. One gives the tension stress, the other the compression.

$$Bending\ stress = \frac{242402 \times 50.65}{320963} = 38N/mm^2$$

This is compression and already slightly above the allowable stress. To this must be added the direct stress, which results in 49 N/mm², much too high.

If this is still a paper design and nothing has been manufactured, then there are a number of changes that can be made. The BM could be reduced by increasing the radius between the front edge of the seat and the leg. Notice that the stress must be reduced by 12N/mm², which is about one-third of the bending stress. This means that the bending arm should reduce similarly, from 329.35 to 220 or less. The radius has to increase to 160mm, so the distance up the chair leg is 219.35. Scaling the stresses, as they are directly proportional to the moment arm:

$$bending\ stress\ becomes = \frac{219.35 \times 38}{329.35} = 25\ N/mm^2$$

Adding the direct stress gives the new total 36N/mm², which is acceptable.

Another change that could be made is to increase the thickness of the legs, but an objection is that this increases the cost of the material and also the weight of the chair. This may not concern you, so as an exercise, work through the section properties table to find a larger I and A, with the thickness at 6 mm or more, and see what you get.

A third change that could be made is to add an extra flange to the section. This would be on the leg opposite to the existing flange. We will make this small to keep it possible to stack the chairs.

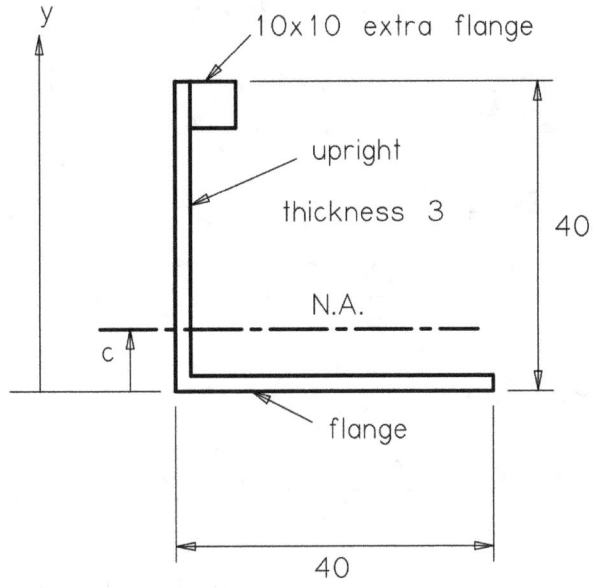

Figure 9.1.5 New section with added flange

New section through leg and seat edge

Item	Width, b	Height, d	y	A=bd	Ay	Ay²	I_s=bd³/12
upright	5	70	35	350	12250	428750	142917
flange	65	5	2.5	325	813	2031	–
Extra flange	10	10	65	100	6500	422500	–
Sum, Σ				775	19563	853281	142917

$$c = \frac{19563}{775} = 25.24mm$$

$$\Sigma Ac^2 = 775 \times 25.24^2 = 493720$$
$$I = 853281 + 142917 - 493720 = 502478mm^4$$

The new bending stress is now

$$\frac{242402 \times (70 - 25.24)}{502478} = 22\,N/mm^2$$

And the direct stress

$$\frac{14720}{2 \times 775} = 10\,N/mm^2$$

The total is now only 32N/mm², which is fine. This is a compression stress; the tension bending stress can easily be found by scaling the compression stress by the ratio of their distances from the neutral axis.

$$tension\ stress = 22 \times \frac{25.24}{70 - 25.24} = 12\ N/mm^2$$

Subtracting the (compressive) direct stress of 10N/mm² gives just 2N/mm² tension.

Another way of reducing the stresses in the legs is by having crossbars which reduces the moment arm. This may become necessary when the legs are splayed outwards to increase the stability of the chair and also make it stackable. The non-vertical angle introduces a horizontal support load, which adds to the moment-induced load described above.

If you are confident of the quality of the material and manufacturer, you may decide to reduce the size of the Safety Factor. This could be dangerous since either or both may change uncontrollably later.

Finally, it might be possible to change materials, and here there are many possibilities which we won't go into, but I hope you will get some clues from this book.

The next item to look at is the arm with someone sitting on it. The load could be assumed to be a UDL (uniformly distributed load). This was calculated earlier as 2.453N/mm before applying a Safety Factor. The maximum BM, which is at the centre, is listed in the reference books as follows:

$$BM = \frac{1}{8}wl^2$$

w is the loading, and l is the length of the arm between supports, which are in this case assumed to be simply supported.

$$BM = \frac{1}{8} \times 2.230 \times 330^2 = 30356 \; Nmm \; unfactored$$

The stress is again given by the usual formula:

$$bending \; stress = M\frac{y}{I} = 30356 \times \frac{(40 - 11.1)}{28513} = 31 \; N/mm^2 \; unfactored$$

The y and I are calculated earlier. As you can see, we need to apply a SF of ten, so this component is seriously under strength. The designer needs to be told to use the same section as suggested for the leg. This gives the following result:

$$bending \; stress = 30356 \times 10 \times \frac{(70 - 25.24)}{502478} = 27 \; N/mm^2$$

The 10 is the factor and this result is fine.

Just a word about buckling in the leg under the compression load, although with such a good section, it is very unlikely:

$$Buckling\ load = \frac{\pi^2 \times EI}{l^2} = \frac{\pi^2 \times 1300 \times 502478}{430^2} = 34800N$$

E=1300N/mm² is from the table in Chapter 8, I and l from the workings above. As can be seen, the load to buckle the leg is enormous compared with the applied loads in this case, so no problem.

Of the beams, it remains to check the front edge of the seat. This is part of the seat pan, of course, but it is usual to analyse it as though it were an isolated component. So we could take Figure 9.1.5 upside down, assuming that the flange is the active part of the seat pan in this beam. The remainder of the seat pan will, of course, help with the bending, but diminishingly so as it recedes from the front edge. This is due to something called shear lag and is also because of the curvature.

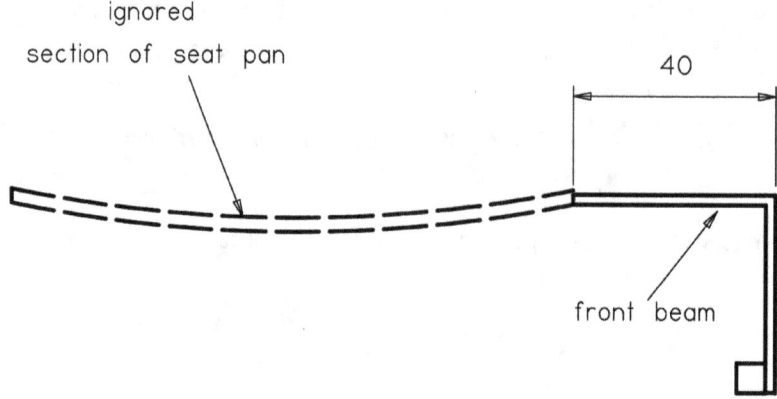

Figure 9.1.6 Section through seat pan from front to back, showing assumed section of front edge beam

The loading from load case 2 is 14720 N (which is factored) spread over the 430 mm width of the chair (this is the length of the beam, l).

The loading is therefore 34.23 N/mm, and the reference books give the bending moment for a SS beam such as the arm:

$$BM = \frac{wl^2}{8} = \frac{34.23 \times 430^2}{8} = 791141 \; Nmm$$

Too high again (compared with 242402Nmm leg moment), but we are also assuming that the legs restrain the bending of the seat edge. If you look at Figure 9.1.4 you will see that the horizontal loads, H, produce a moment at the joint of leg and seat edge. So we have a beam with a distributed load and a moment at each end, which reduces the BM in the centre of the edge beam.

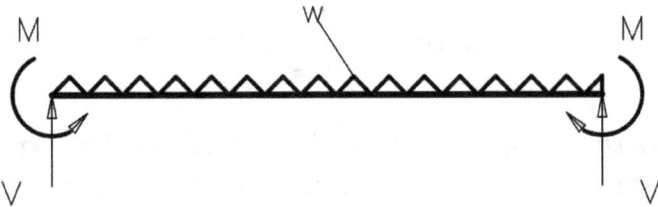

Figure 9.1.7. Seat edge as a beam with UDL supported by vertical load and end moments

The BM at the centre due to both the end moments, is constant along the entire length of the beam.

$$BM = Hl = \frac{1}{20}wl^2$$

The end moment is reaction load H, as calculated before, times the distance up the leg. Examining Figure 9.1.7, you can see that these moments tend to bend the seat edge up, whereas the loading, w, bends the seat down. In other words, they subtract one from the other.

$$Net \; BM = \frac{1}{8}wl^2 - \frac{1}{20}wl^2 = \left(\frac{5}{40} - \frac{2}{40}\right)wl^2$$

$$= \frac{3}{40}wl^2$$

So we have dropped from five fortieths to three fortieths of the BM by accounting for the effect of the legs being pinned at their feet. This is more accurate but somewhat more complex; when the initial simple analysis gives too high a stress, it is often true that a more complex one will give a lower stress. If this is more than you are prepared to undertake, you should ask an expert; he may use his Finite Element computer program to analyse the entire chair quickly and in detail. It all depends on your objective.

The BM reduces to 60% of the SS beam one i.e. to 474685 Nmm. This is still more than the arm BM. The bending stress is 42 N/mm². There is a direct stress due to *H.*

$$direct\ stress = \frac{736}{775} = 1\ N/mm^2$$

This is compression, and the total is therefore 43 N/mm² compression. To get the stress down further you could increase the size of the extra flange. Try it with this measuring 20 x 20, as a starter.

This analysis has assumed that the leg length and seat width are equal (and that their beam Is are equal) in order to reduce the complexity of the equations given by the books. If they are within 10 per cent of each other, the stresses vary less than 15 per cent from our case. If the front edge's I is larger than the leg I, then the value of H falls so the supporting moments reduce. If the edge I is twice the leg I, the support moments reduce to 71 per cent of the original; of course, the bending stress the beam can carry will double. The leg maximum BM will also reduce. Again this comes from studying the equations in *Roark.*

Another case which may give a high BM in the front edge is case 4, that of a man standing at its middle, which indeed it does.

It is again made up of the BM due to the load, which is the point load in the middle of the seat edge beam considered simply supported, plus the moments due to the horizontal loads at the bottom of the legs. This is the same as in Figure 9.1.7 except that the load is a central point load. The maximum moment, at the centre, due to this point load is

$$M = \frac{Pl}{4}$$

P is the load, which is the man standing on one tiptoe. The horizontal load, making our assumption of equal *I* and length of seat front beam and leg, is as follows:

$$H = \frac{3P}{40}$$

So its BM is

$$M = Hl = \frac{3Pl}{40}$$

As before this is constant over the seat edge because there is a moment each end.

The total BM is the difference of the two

$$M_t = \frac{Pl}{4} - \frac{3Pl}{40} = Pl\left(\frac{10-3}{40}\right) = \frac{7}{40}Pl$$

If we notice that the Pl is equivalent to the wl^2 in the previous case, then we can also see that this BM is over twice as big. This makes the bending stresses also over twice as big. Again, there are a number of ways of dealing with the situation. Would a 150 kg man clamber onto a light garden chair? You might say that only a 75 kg chap would. The heavier man should know from experience not to do such a thing. Then again, he might not, so if none of the other fixes – reducing the SF, increasing general thickness, and so forth – appeal, the section I will have to be increased, another practice opportunity for you.

The last component we will consider is the seat pan. The design of this can vary; it could be continuous – that is unpierced – or it could have holes or slots for rain drainage, for instance. Taking the continuous one, we earlier got a pressure of 0.02074 N/mm^2 unfactored.

The seats of plastic garden chairs are usually slightly curved. Even if they are flat when manufactured, they are usually thin and will therefore bend into a curve when a heavy load is applied. The pressure quoted above is from case 2, a heavy man relaxing suddenly onto the chair. The load will be reacted by tension in the seat pan rather like a cylinder under pressure. The stress therefore is

$$cylindrical\ or\ hoop\ stress = \frac{pR}{t}$$

We know the pressure, p, and thickness, t, but R must be measured. We will use a geometry theorem concerning chords to calculate it. In Figure 9.1.8(A) the theorem says that the multiple of the lengths of each part of any chord is equal to the same multiple of any other chord:

$$a \times c = b \times d$$

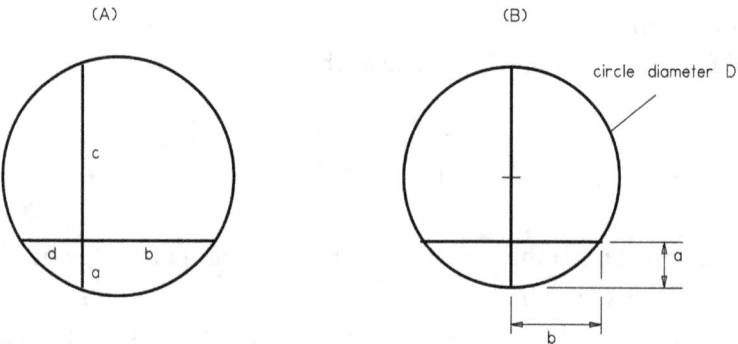

Figure 9.1.8. Intercepting chord theorem, ac=bd

If we draw one chord through the centre, as in Figure 9.1.8 (B), it obviously bisects the other chord and then

$$(D - a) \times a = b^2$$

$$D = \frac{b^2}{a} + a$$

A cross-section of the seat pan is represented by the lower arc in (B).

Now take a straight edge and a ruler and place the straight edge over the seat pan. Measure the distance from the deepest part of the pan upwards to the straight edge. This is *a*. Also measure from the front to the back of the pan. This is *2b*. *D* can now be calculated.

In our case, *a* is 15mm, *2b* is 330mm, and *D* is 1830mm. The radius is half this, 915mm.

The hoop stress can now be calculated.

$$hoop\ stress = \frac{pR}{t} = \frac{0.02074 \times 10 \times 915}{5} = 38\ N/mm^2$$

The equation, for Case 2, includes the SF of 10.

This assumes that the seat pan is cylindrical in shape, so the load is delivered to the front and back edges only or the side edges only. If the pan is spherically dished, the load will be reacted by all four edges and the stress in the pan is halved to 19N/mm2 because the hoop stress in a sphere is half that in a cylinder. Clearly we must do this. Note that this load also bends the seat edges sideways.

I have left the fatigue case to to the end, as the calculations are somewhat more extensive. It is first necessary to calculate the stresses as above but unfactored. The relevant stress is always tensile, as fatigue occurs only in this state. The loads are due to the child's mass of 25 kg. We will assume for this exercise that we can just reduce the stresses we have found in the various components by the ratio of the child's mass to that of the man's mass. This would be exactly accurate for metals but not quite so for unreinforced plastic, a point we have ignored. The reason is that the load to deflection relationship is not quite linear for the plastic. In most circumstances, though, the results are good enough.

It is desirable that the stress is less than the long-term fatigue allowable, or limit stress. If it is not, we would have to find a stress-to-life curve for the material. In our case, we have assumed polypropylene and a fatigue limit stress of 20 per cent of the ultimate of 37N/mm², which is 7.4N/mm². This is at 10^7 cycles.

Examining our stresses above for the various parts in the various cases, we find the worst tensile stress is 3.8N/mm² unfactored in the seat pan if there are only beams at front and back of the pan. This is for the 150 kg man; for the 25 kg child, it is one-sixth of that, 25/150, which is 0.63N/mm². We can mark this on the (diagrammatic) figure below.

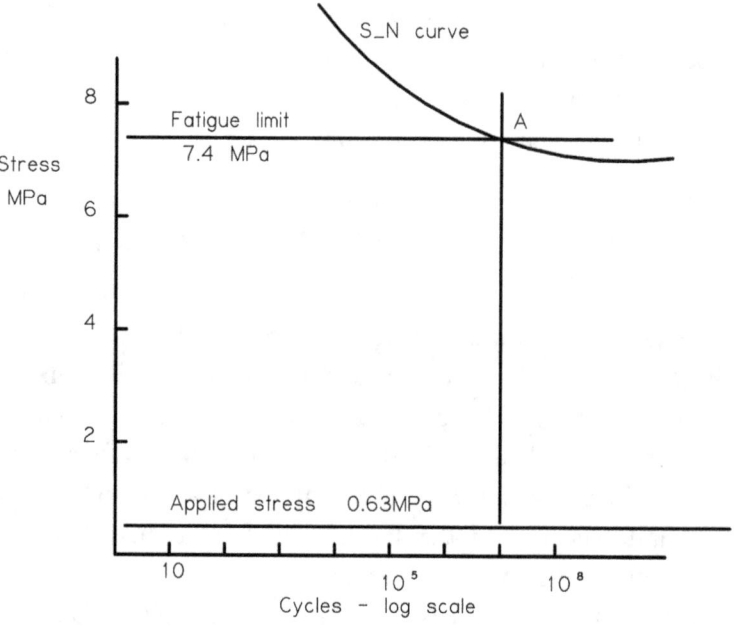

Figure 9.1.9. Diagrammatic S-N curve with fatigue limit at 10⁷cycles and applied stress

I have shown the S-N curve as typical, and it should not be used other than in the explanation. Point A shows our fatigue (or endurance) limit, and it is obvious that our child can jiggle forever because the applied stress in this case is way below the limit. If the stress had been above the limit and we knew of a properly derived S-N curve, we would be able to read off the cycles to failure. This would then be divided by the SF of five to give the allowable life.

Ch. 8 Section (d) Comparison and RF

The last thing to do is calculate the Reserve Factors and list them in an orderly fashion so that it is easy to see each item that has been considered and its location. The best thing is to construct a table of the results. The columns can contain various headings. Essential ones are the name of the component, the position and type of the stress, the values of the allowable and applied stresses and their quotient, the RF, according to the equation

$$RF = \frac{allowable\ stress}{applied\ stress}$$

Other quantities to be specified are the material of the component, drawing numbers if they exist and you have them; a comment column is also useful.

RFs for the garden chair

Material is polypropylene; stress units N/mm^2

Item	Position	Stress type	Allowable stress	Applied stress	RF	comment
Leg	Top under radius	Bending and direct compression	37	32	1.16	With modified section
Arm	Centre	Bending	37	27	1.37	With new section
Seat front beam	Centre	Bending and compression	37	<37	1.0 +	Using your newest section
Pan	All	Tension	37	19	1.95	With all round edge beams
		Fatigue	7.4	0.63	–	Life is infinite

From this list, the RF column makes it obvious where the critical part of the chair is. Clearly, in a more complicated structure, the list would be longer and may include bolted joints, welds, and so on. Anything that is relevant to the strength of the structure can be incorporated.

2. The fan

Ch 1   Load cases

As another example, consider a fan heater. It gets hot, the fan rotates, and there may be vibration problems. There are failure and misuse cases to think about. It will probably be constructed of different materials – metal for the fan blades, perhaps, and maybe plastic for the casing.

To describe the rotational loads you need to discover the speed of rotation and the mass and shape of the blades. There is also the case of bending the blade due to the fluid pressure on it. This will add to the centrifugal force. To find the effect of the temperature, you need to know its distribution – that is, how it varies in the different parts of the machine. For failure cases, you study the detail of the construction and think what parts may break and the effect of that. Similarly, for misuse cases, you may decide that someone may stand on it. It should be able to resist that successfully, or you may design it so that it is uncomfortable or impossible to stand on. Then there is no load case.

Ch 2   SF

Again, there seems no point in low RFs. So use the same ones as in the previous example and for the same reasons. The rotating fan blades need to be safely anchored to the hub, so it may be sensible to use a higher RF, perhaps 2 for yield and 3 for ultimate for metals. For plastic blades, 10 should give a low working stress.

Ch 3   Structural data

As usual, everything must be measured. As the stress in the rotating blade may be a critical feature, its dimensions must be accurately established, including its thickness, which may not be uniform. This is because its mass must be found; alternatively and better, if you have a blade handy, you can weigh it. You can also balance it on a pencil and find its centre of gravity.

The materials may well vary considerably in different parts of the fan. The casing could be steel and the blades plastic, or vice versa. The

234

motor and electrical switchgear will be proprietary items which need not be stressed.

Ch 4   Classification by component type

The worst stress is going to be in the roots of the blade due to the centrifugal force or bending due to the aerodynamic loading. This means that the blade is either a tension member, whatever its cross-sectional shape, or a bending member, or possibly both. If the fan casing is of such a shape, roughly cubic, that someone may use it to stand on, perhaps to reach a high shelf, then that is a load case and the top surface will be a plate loaded in bending. The sides supporting the top will be in compression.

Ch 5   Support reactions

Again, not much can be said except that the fixing supports the weight of the fan. Only if the fan is of industrial strength will the thrust of the air be significant, and this gives an overturning moment which will add to the load on the supports. This can be calculated in a similar manner to the example shown in Figure 5.9, by adding the overturning moment's reactions to that of the weight.

Ch 6   Loads and their application

The most important and highly stressed part is the blade. If it is a high-speed fan, the largest load will be the centrifugal load; if it is slow and in water, then the highest stresses will come from the pressure over the blade, which will bend it at the root.

The centrifugal load is given by the expression

$$Mr\omega^2 \text{ Newtons}$$

here M is the mass of the blade in kg, r is the radius of the centre of gravity of the blade in m, and ω is the rotational velocity in radians per second. There are 2π radians in 360 degrees, or 57.3 degrees per radian. This gives the tension load at the blade root.

The bending due to the pressure over the blade is more difficult because the pressure varies not only along the blade span but also across the chord. As an approximate estimate of an equivalent point load, assume it is centred three-quarters up the span and a quarter of the way from the leading edge. The centre of pressure may be difficult to calculate if the blade is an irregular shape; often blades widen with increasing span. They may also be bent and twisted, meaning experts may have to be consulted. The load is

$$\tfrac{1}{2}\rho v^2 S C \text{ in Newtons}$$

Here, ρ is the density of the fluid in kg/m^3, v the linear velocity of the blade at the centre of pressure in m/sec, and it equals $r \times \omega$. S is the area of the blade in m^2. C is a coefficient of lift. If eventually this approximation ends up giving a high stress, then a better calculation needs to be made by an expert in the field.

It may be necessary to find the temperature distribution over the hot parts of the fan since varying strains may develop in the hot and cold parts.

Ch 7   Calculations

For our detailed calculation example, let us consider a simple small cooling fan. It has a rectangular blade which is of uniform thickness. This makes it easy to calculate its mass and centre of gravity (CG). We will concentrate on stressing the blade, as the loads on the casing and/or the supports will be much the same as for the chair and can be similarly stressed.

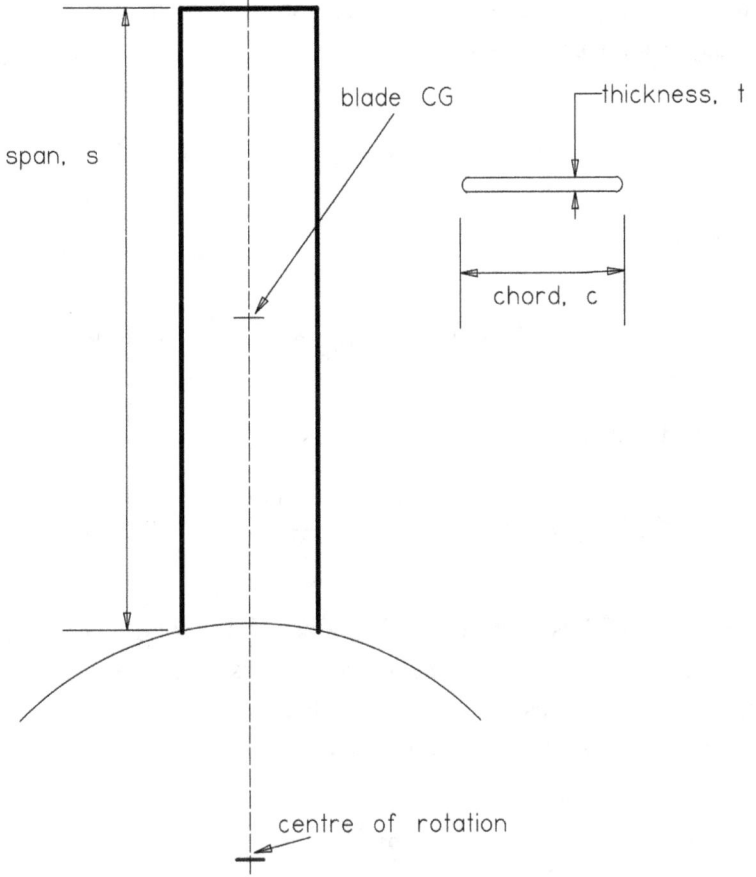

Figure 9.2.1 A simple fan blade

In Figure 9.2.1, we will take the dimensions as being s=70 mm, c=20 mm, t=1 mm and the radius of the centre, and CG, of the blade to be 63 mm. The mass of the blade, assuming it is made of aluminium with a density of 2.7 gramme per cubic centimetre or 0.0027g/mm³ is

$$M = sct\rho$$

$$M = 70 \times 20 \times 1 \times 0.0027 = 3.78g$$

This is 3.78 10⁻³kg. The radius we have, the other thing we need, is ω, the rotational speed in radians per second. As mentioned, there are 2π

237

radians for every revolution and, of course, sixty seconds per minute, so at, say, 3600 rpm, the value is

$$\omega = \frac{2\pi n}{60} = \frac{2 \times \pi \times 3600}{60} = 377.0\,^{rad}/_{sec}$$

The centrifugal force is

$$M r \omega^2 = 3.78 \times 10^{-3} \times \frac{63}{1000} \times 377.0^2 = 33.85N$$

The direct tension stress is low in our case:

$$centrifugal\ stress = \frac{force}{area} = \frac{33.85}{20 \times 1} = 1.69N/mm^2$$

It seems we could use a pure plastic material based on this result.

The second loading case must also be considered; this is the fluid load. This will obviously add to the CF load.

The load equation, as explained above, is

$$\frac{1}{2}\rho v^2 SC$$

v is a linear velocity equal to $r\omega$ – that is, it is the tangential component and it varies up the blade; the tip goes fastest and the root slowest. The variation in the load is not linear because of the squaring of the velocity. The result of this is that more of the load is at the tip so the root BM is higher than if the variation were linear. Therefore, we cannot make the assumption that the centroid of the load is halfway up the blade; that would underestimate the root BM, a non-conservative assumption. A conservative assumption is that the centroid is three-quarters of the span up the blade from the root.

The density of air is 1.25kg/m³. This is at what is called the standard temperature and pressure (STP) – that is, 15°C and 1 bar (14.7psi).

In our example, ω is 377 radians per second and the radius of the blade varies between 28 mm and 98 mm.

S is the area of the blade, which is 20 mm times 70 mm, or 1400 mm². In fact, it needs to be in m², so we have 1400×10⁻⁶m².

The coefficient of lift, C, varies according to the shape of the blade, its curvature in the direction of fluid flow, the angle of incidence to the flow, and other parameters. For our example – a flat rectangular plate at a small angle of incidence – it is equal to $2\pi a$, where a is the angle of attack in radians. We will take it as 1.0 for simplicity, and this is not a bad estimate. The load is as follows:

$$\tfrac{1}{2} \times 1.25 \times 377^2 \times 1400 \times 10^{-6} = 124.4N$$

We are taking the centroid of the load as three-quarters of the way up the blade, which is

$$\frac{3}{4} \times 70 = 52.5mm$$

The BM at the root is

$$124.4 \times 52.5 = 6531Nmm$$

The second moment of area, I:

$$I = \frac{1}{12} \times width \times thickness\ cubed$$

This is a standard result for bending a rectangular bar:

$$I = \frac{1}{12}st^3 = \frac{1}{12} \times 20 \times 1.0^3 = 1.667mm^4$$

As usual, the stress is

$$stress = M\frac{y}{I} = 6531\frac{0.5}{1.667} = 1959\ N/mm^2$$

This is obviously far too much, and we haven't applied a Safety Factor yet. What to do depends on the circumstances. You can reduce the speed to a safe level. We notice that the load is proportional to the speed squared. This means we can reduce the speed by the square root of the ratio of the allowable stress to the applied stress, the stress being proportional to the load.

If we take a normal Al alloy and consider the allowable stresses with their Safety Factors (see Chapter 8, section (b)), we find:

Ultimate allowable 310N/mm² with a SF = 2 gives a working max of 310/2=155.

Yield allowable 260 and a SF of 1.5 gives a working max of 173.

Fatigue limit of 60 per cent of 310 gives a working max of 186.

The lowest maximum working stress allowed is in the ultimate case. The allowable rpm is therefore:

$$\sqrt{\frac{155}{1959}} \times 3600 = 1013 rpm$$

If you are designing from scratch, then you know how fast this fan can be allowed to rotate. If you have an existing fan and you wonder whether you can increase its speed, you now know. It should be noted that if the SFs reduce to 1.5 ultimate and 1.25 yield, then the critical allowable stress is now the fatigue limit, and this results in an allowable speed of 1109 rpm. These SFs, however, need reliable manufacturing and so on, so they should be used cautiously.

Another fix is to increase the thickness of the blade. This increases the CF load linearly, but we have found that it is small. It decreases the root bending stress by the square of the thickness. This is because the I is proportional to the thickness cubed, but the stress is increased by the thickness, as in the following:

$$stress = \frac{My}{I} = \frac{M\frac{t}{2}}{\frac{st^3}{12}} = \frac{Mt}{2} \times \frac{12}{st^3} = \frac{6M}{st^2}$$

s is the span, *t* is the thickness, and as before and *y* is half the thickness.

We found that the bending stress is 1959N/mm^2 for a 1 mm thick blade, and this obviously needs to be reduced by about an order. If we increase the thickness to 3 mm, the stress will come down by a factor of 3^2 or 9, to become 218N/mm^2. The CF stress becomes 1.69 x 3, which is 5N/mm^2, so the total new maximum stress is 223N/mm^2. This is not enough against the allowable of 155N/mm^2 ultimate, but obviously of the right order. To find the minimum thickness, we must scale it:

$$\min t = \sqrt{\frac{1959}{155}} \times 1 = 3.56mm$$

Adding 6N/mm^2 for increased CF stress means that at least 3.6 mm is required; in practice, the blade will probably be cut from a 4 mm plate, which is a standard size.

If the blade can be made curved at an acceptable cost it will come out thinner as you get a better *I* for a given thickness. Another alternative is to taper the thickness so that it is thicker at the root. It would be expensive to produce a continuous taper, unless the blade is cast, so maybe short supporting pieces of plate could be used on one or both sides of the blade.

An unreinforced plastic blade does not look practical, but a fibre-reinforced one does, provided there is plenty of reinforcement in the radial direction.

Blades that are of a more complicated shape because they deliver more fluid flow for a given size or power will need more detailed analysis. If they are twisted up the span, they may develop torsion at the root. This is because the centrifugal force is not balanced. All this should be left to specialists.

Notice the method of calculating the effects of the changes: scaling of the preliminary result by our alterations to get the altered stresses. This shortcut can often be used; if you are not used to it or comfortable with it, then you can do the formal calculations each time. If you have used it extensively to arrive at a final answer, it is worth doing the formal detailed calculation with your final parameters as a check on your workings.

3. The joist

Ch 1   Load cases

If you wanted to bring a particularly heavy machine or other item into an upstairs floor of a building and it was clear that the floor would have to be reinforced because it was already heavily loaded, or the building was old and the floor suspiciously bouncy, you might decide to insert an extra floor beam. In practice, there will be more than one joist to support a particularly heavy piece of kit. We will assume that all will be loaded equally so only one needs designing. Or, if it were found that one was loaded more heavily than the others, then that would be designed. The others would probably be made the same for simplicity.

If the building is modern, it will have already have been stressed to a point load in order to conform with the Building Regulations. The size of this load is dependent on purpose of the building; it varies between 1.4 kN and 9.0 kN. This is not actually a point load; it can be spread over a 300 mm square. By the way, roughly it's 10 kN to the tonne.

This case is obviously dominated by the weight of the item to be supported on the floor; care must be taken, however. If it's a machine, what if it vibrates? If it has a heavy moving part in it, the supporting legs may have a varying load in them. There may be a failure situation causing unusual loading. The other load types do not seem to cause load cases.

Ch 2   SF

The Building Regulations specify a "design stress". As such, no SF is used. Instead, this is incorporated in the design stress. The work of providing a safety margin is done for you. Of course, if the use is not controlled by the Regulations then a suitable SF must be chosen according to circumstances. These will be 1.5 and 2, or 2 and 3, for yield and ultimate, and 2 on life if the stress is below the fatigue limit.

Ch 3   Structural data

This should be easy to measure; the beam will usually be of a standard size and grade, and its properties can be looked up in structural handbooks for steel, aluminium, or timber. Special one-off beams will have to be specified individually.

Ch 4   Classification by component type

This is clearly just a beam.

Ch 5   Support reactions

Look at Figure 9.3.1. This is a simply supported beam with various loadings. If there are more than two supports the situation is more complicated because it is redundant, but this one is OK.

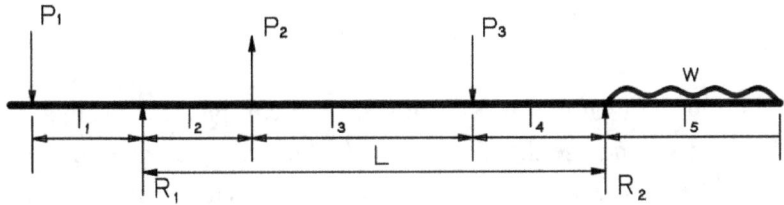

Figure 9.3.1 Beam with overhangs at each end and various loadings

Taking moments about support R_1 gives us R_2 directly; load and moment directions as in the Figure.

$$R_2 L = -P_1 \times l_1 - P_2 \times l_2 + P_3(l_2 + l_3) + w \times l_5 \times \left(l_2 + l_3 + l_4 + {}^{l_5}/_2 \right)$$

Dividing the RHS by L gives R_2; probably too lengthy a calculation to do on the calculator in one go, especially if the numbers are complicated.

Do each bit separately and add up after that. Vertical balance then gives R_1.

$$R_1 = P_1 - P_2 + P_3 + w \times l_5 - R_2$$

Ch 6   Loads and their application

Floor joists usually only carry two types of loads: concentrated or distributed loads. In buildings, these are specified in the Regulations according to the use made of the building. A change of use may mean the floor's strength needs to be re-evaluated. It may be possible to do this from the tables of sizes of joists required, as given in the Building Regulations. If not, special calculations on the most heavily load joist will have to be carried out. The relevant equations giving the maximum moments and shears can be found in the usual books.

If the joist is not to be used in a floor – perhaps it is a reconditioned one from a demolished building – then it must be considered a simple beam and ordinary calculations can be done. Of course, if it is a reused one, it must be physically examined to ensure it is not damaged.

Chapter 7   Calculations

Taking the example in Figure 9.3.1, we can put some dimensions and loads to it and use that for some calculations. First we will stress it for static loads, then assume the loads are from a vibrating machine and consider the fatigue case. Finally, we will do some simple dynamics.

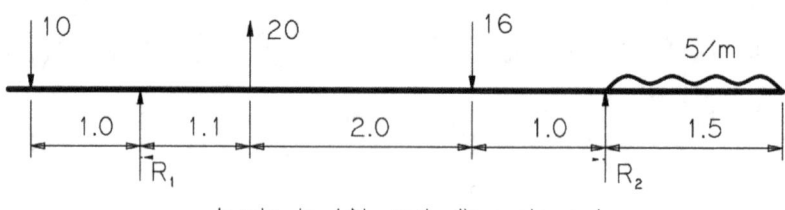

loads in kN and dimensions in m

Figure 9.3.2 Loads on the beam

Using the general equation above, we get the following for R_2:

$$R_2 \times 4.1 = -10 \times 1.0 - 20 \times 1.1 + 16(1.1 + 2.0) + 5 \times 1.5\left(1.1 + 2.0 + 1.0 + {}^{1.5}/_2\right)$$

$$R_2 \times 4.1 = -10 - 22 + 49.6 + 36.375$$

$$R_2 = \frac{53.975}{4.1} = 13.165$$

$$R_1 = 10 - 20 + 16 + 5 \times 1.5 - 13.165 = 0.335$$

The best way to check your answer is to take moments about R_2. Try it.

The next thing to do for this beam is to draw the shear and BM diagrams. This will indicate the positions of the maximum shear and BM and therefore the maximum stresses.

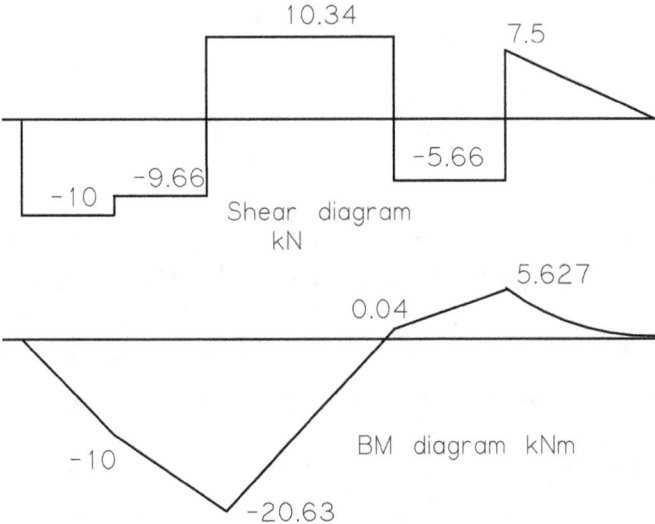

Figure 9.3.3 The shear and BM diagram for the beam in Figure 9.3.2

The shear and BM diagrams are constructed as described in Chapter 7 section (c)(2); the maximum shear is 10.34kN. Incidentally, if you need a joint in the beam, it is obviously best to put it between P_3 and the left support since the shear and BM are low.

The maximum BM is -20.637 kNm, with the minus sign meaning that the top is in tension in this example. This is obvious if you look at the cantilevers at either end. The loads on them bend the cantilevers into downwards arcs, causing the top to stretch. In some industries a convention is followed that the negative sign always means compression and a positive one means tension, but this requires that a strict procedure is followed. I have not done this as it is usually obvious in simple examples what is in tension and what not.

If the beam is an old-fashioned joist three inches deep, we can look up its cross-sectional properties, its I and A, weight per metre, and so forth, in the reference books or online.

Switching to metric dimensions, a 76x76x13 mm joist has an I = 1.58×10^6 mm^4. The maximum bending stress is the usual

$$\frac{M y}{I}$$

Note that there are 10^6 Nmm in a kNm.

$$\frac{20.63 \times 10^6}{1.58 \times 10^6} \times \frac{76}{2} = 496 \,^N/_{mm^2}$$

The steel is BS 570 M44, with ultimate strength 850N/mm^2 and yield 630N/mm^2. Its fatigue limit is 50 per cent of ultimate (that is, 425N/mm^2).

In this case, use an ultimate Safety Factor of 1.5 and a proof or yield one of 1.25. So the factored ultimate is 1.5x496 or 744N/mm^2, which is less than the allowable of 850N/mm^2, giving a Reserve Factor (RF) of 850/744 = 1.14.

For yield, the factored stress is 620N/mm^2 and the RF is 1.02. Both these RFs are fine. But look at the fatigue limit – it's only 425N/mm^2. If the loading represents some sort of oscillating load, there is a danger that fatigue failure may occur. Remember, though, that the alternating part of the stress is most important; we have no information on this currently.

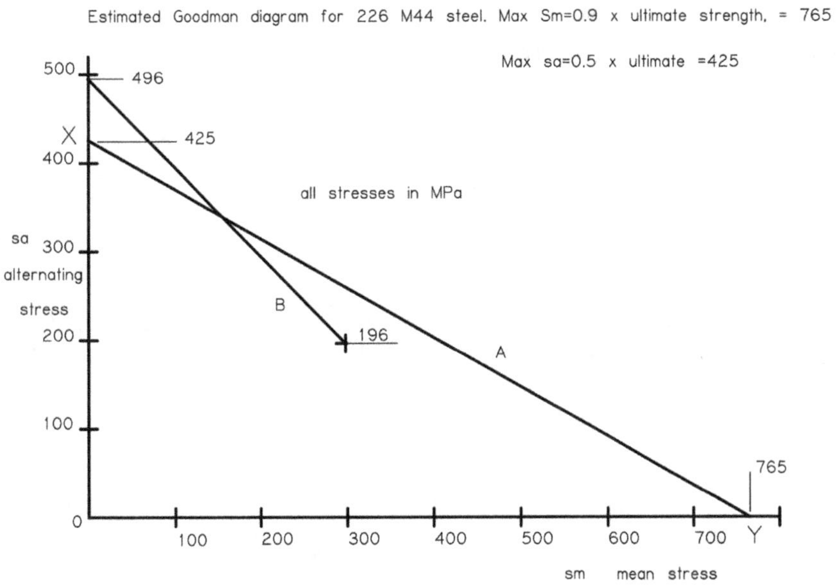

Estimated Goodman diagram for 226 M44 steel. Max Sm=0.9 x ultimate strength, = 765

Figure 9.3.4 Goodman diagram for the steel in the example with the applied stress

Look at Figure 9.3.4. Line A is the Goodman line, as described earlier. It stretches from point X on the vertical axis (or the alternating stress axis), which indicates zero mean stress, and the 425Mpa fatigue limit stress, to point Y on the horizontal axis (or the mean stress axis), which indicates zero alternating stress and 0.9 times the ultimate stress. Below this line, the applied stress is satisfactory; above it, fatigue may occur.

Line B stretches from our calculated maximum stress to another point I have chosen, where the mean stress is 300Mpa and the alternating stress is our maximum minus this 300MPa, which is 196 MPa. This means that the minimum stress is 104Mpa and the point is a safe point. The lines cross at about a mean stress of 158Mpa and an alternating stress of 338Mpa.

In the case that fatigue is an issue and an estimate of life is required, it is necessary to have an S-N curve (the number of stress cycles for a range of alternating values of stress at failure). These are usually obtained by testing numbers of samples of the material, as described earlier. However, if such data is not available, a theoretical S-N curve can be constructed for wrought steels.

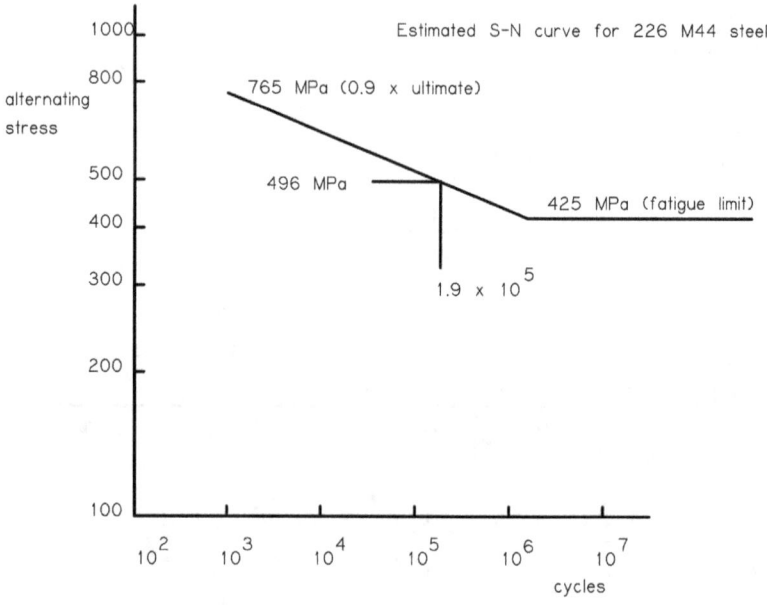

Figure 9.3.5 A theoretical S-N curve for 226 M44 steel

This fatigue curve starts at 10^3 cycles and 0.9 x the ultimate strength and finishes at 10^6 and half the ultimate. This is the usual procedure for wrought steel of ultimate strength of 1400Mpa or less; above that strength, the assumption is that the fatigue limit remains at 700Mpa. This is a conservative curve (as it ought to be) and should only be used in the absence of better data.

Now, if the loading on our beam is completely reversed, then we have the case of zero mean stress, and the alternating stress is the maximum bending stress we have found (496Mpa). This is the most severe case and is marked out on Figure 9.3.5. It gives a life of 1.9×10^5 cycles. Using the SF on life of five, this means a usable life is 38000 cycles. Of course, it all depends what causes the alternating load as to whether this is acceptable. If the situation occurs once a day, there will no trouble for one hundred years, but if it's once a second, you may have little more than ten hours of life.

If the loading oscillates at anything like the resonant frequency of the beam, then the deflections in it and therefore the stresses will increase, which is unacceptable. In Chapter 7 section (e), it was pointed out that an easy way to get an estimate of the natural frequency of a single degree of freedom system is to use the equation

$$f = \frac{1}{2\pi}\sqrt{\frac{g}{d}}$$

f is the fundamental natural frequency, g the acceleration due to gravity, d the deflection.

To demonstrate how this works, we will take a simple example of a beam with the same geometrical properties as the one above but with only a simple point load at its centre.

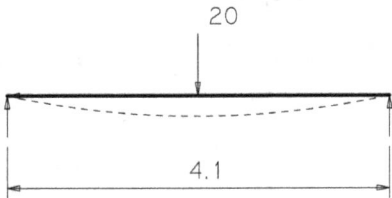

Figure 9.3.6 Simply supported beam with a load of 20 kN and length of 4.1 m

The central deflection is

$$d = \frac{PL^3}{48EI} = \frac{20000 \times 4100^3}{48 \times 200000 \times 1.58 \times 10^6} = 90.88mm$$

Note that everything has been converted to Newtons and mm. We now put this deflection into the frequency equation.

$$f = \frac{1}{2\pi}\sqrt{\frac{g}{d}} = \frac{1}{2\pi}\sqrt{\frac{9.81}{90.88 \times 10^{-3}}} = 1.653 \; Hertz$$

Note that here the equation demands metres so the deflection had to be converted.

The answer is typical for big stiff structures and might well be near enough as an estimate of the fundamental mode for first investigations.

Our beam is not a single degree system, but the equation will give a first estimate to within perhaps 10 to 15 per cent. The job now is to calculate the deflection.

Examining Figures 9.3.1 and 9.3.3, it is easy to see that the deflection of the beam is due to the four loads; the overhanging loads can be converted to moments at each end (equal to $P_1 \times l_1$ and $w \times l_5^2/2$) in order to calculate the deflection in the middle span of the structure. The usual sources will provide equations which calculate the deflection anywhere along the beam for each load. These deflections can be added together to give the total. The question is exactly where. The BM diagram in Figure 9.3.3 shows a maximum at 1.1 m from the left-hand support. Is this the point of maximum deflection? No, it is not, although it won't be far away. I constructed a spreadsheet to find this total deflection.

Deflection at x m due to M_1, where x is anywhere along the beam between R_1 and R_2:

$$defl = \frac{M_1}{6EI}\left(3x^2 - \frac{x^3}{L} - 2Lx\right)$$

Deflection due to M_2:

$$defl = \frac{M_2}{6EI}\left(Lx - \frac{x^3}{L}\right)$$

Deflection at load P_2 due to P_2:

$$defl = \frac{P_2}{3EIL} \times a^2 \times b^2$$

a and b refer to the position of the load, a is 1.1 and b is 3.

Deflection at load P_2 due to P_3:

$$defl = \frac{P_3}{6EIL} \times bx(2L(L-x) - b^2 - (L-x)^2)$$

In this instance b is 1.0 as it is the position of P_3.

The reason I have not worked through each equation in small steps is that I leave it to you to do this. Extract the EI term and apply it in one go at the end and do each part of each equation separately, writing down the answers as you go. This makes it less likely you will make a mistake and easier to find it if you do. Remember, the equations are all in m and kN so; the answers will be in the same units.

I used a spreadsheet as this avoids arithmetical errors. Computers don't make them. The other advantage of a spreadsheet is that it can be used to find the maximum deflections relatively easily. Once you have entered the equations correctly, you need only change x and the new deflection is instantly found for you. Advancing along the beam, the deflection increases up to a point and then reduces. By decreasing the interval in x, about the maximum found so far, the deflection can be found ever more accurately. The maximum I found by this method is 72.98 mm at $x = 1.65$ m. This is 11.35 per cent higher than at the point of maximum stress.

Back to dynamics, at the maximum deflection point, the frequency is

$$f = \frac{1}{2\pi} \sqrt{\frac{9.81}{0.07298}} = 1.845Hz$$

Note that the deflection must be in metres because g is. This frequency is 10 per cent bigger than the frequency of our simple example. For our purposes, this is not a large difference.

So the fundamental natural frequency is in the region of 2Hz, and there will be higher modes of the same order. If there might be a problem with your structure, remember that this analysis is of the beam only. The supports may also be soft enough to contribute to the vibration. Seek expert help.

The oscillations of the loading used in these calculations should be nowhere near these frequencies. If they are, expert computerised help should be sought about this too, in order to get better estimates.

4. The wheelie bin

Ch 1   Load cases

Say you wanted to move a tank of water or other liquid about for some reason but did not want to spend much money. Your eyes light upon the wheelie bin, which you decide to fill with this liquid so that you can move it around. You can measure the bin, calculate the volume, and determine the weight of water. Cases to consider are mass, the pressure of the water on the flat sides near the bottom and the bottom itself, vibration as it is pulled over rough surfaces, possible failures, and misuse.

Ch 2   SF

These bins are made of unreinforced plastic supported on two plastic wheels. The axle is a steel tube. Obviously, the three items need to be considered separately. For the bin, use a factor of ten for safety-related use; the liquid might be damaging – say, acid. For the axle, pick the SFs from the table as usual.

The wheels will be proprietary items, and the manufacturer's catalogue will give the safe load. He should have incorporated the statutory Safety Factor. But you may need new wheels if the weight is excessive – remember, these bins cope with being full of newspaper when used for recycling, so they are reasonably strong.

Ch 3   Structural data

Getting the dimensions, including the material thicknesses, should be straightforward. Finding out what the material is is more difficult. It could be polypropylene. The manufacturers should be consulted; failing that, a conservative estimate of the strengths will have to be assumed.

Ch 4   Classification by component type

With the bin full of a liquid, the sides are plates in bending; there will be tension as well because the sides will tend to act as membranes. Another component to consider is the axle which is a beam in bending.

Ch 5   Support reactions

A bin full of a liquid has two load cases. The first is when it is standing upright on the ground, and the second if it is to be lifted by the handle.

In the second case, lifting a bin full of water is likely to break the existing handle, which is merely plastic. Either abandon the case – never lift it when full by the handle – or if a suitable modification is possible, the reaction to the weight of water and bin is just that. The angle of the reaction will vary with the tilt.

Ch 6   Loads and their application

In a bin full of liquid, the two obvious places that need attention are the axle and the sides and bottom of the bin. The axle is a beam and can be dealt with as in Example 3.

The sides of the bin are under pressure, which is treated as a load distributed over the surface. In this case, the pressure is not uniform, as it is dependent on the depth. The equation for the pressure at a depth of h m is

$$p = \rho g h \ \text{N/m}^2$$

Herein, ρ is the density of the liquid in kg/m^3 and g is the acceleration due to gravity. The pressure is directly proportional to the depth of the liquid, so it increases linearly with the depth.

If the bin is filled with water to a depth of one metre, this gives a maximum pressure of the following:

$$pressure = \frac{1000kg}{m^3} \times \frac{9.81m}{sec^2} \times 1.0m = \frac{9810kg}{msec^2}$$

$$now\ a\ Newton, \qquad N = \frac{kgm}{sec^2}$$

Or

$$kg = \frac{Nsec^2}{m}$$

$$so\ that\ pressure = 9810\frac{N}{m^2} = Pa$$

This is substituting for *kg* in the *pressure* equation. A Pascal is a small unit; the usual is MPa so that the *pressure* becomes 0.009810 MPa.

See table in Chapter 7, section (b) for units of pressure. This pressure acts across the whole of the base of the bin but decreasingly up the sides.

Chapter 7   Calculations

Figure 9.4.1 below shows sections through a typical wheelie bin. It is full of water. Under the pressure of the water, the sides will bulge out a little under bending and tension because each side will act as a membrane, which means the material will stretch. There are, therefore, two load paths. Initially, the plates will bend as plates between the corners, but the tension in the plates will limit the bending deflection. How do we estimate what proportion of the load is carried by each load path? The first thing to do is calculate the central deflection for each path, putting the full load on it alone. We know the tension stress from statics and can therefore work out the deflection for the membrane load path. The bending deflection can also be calculated.

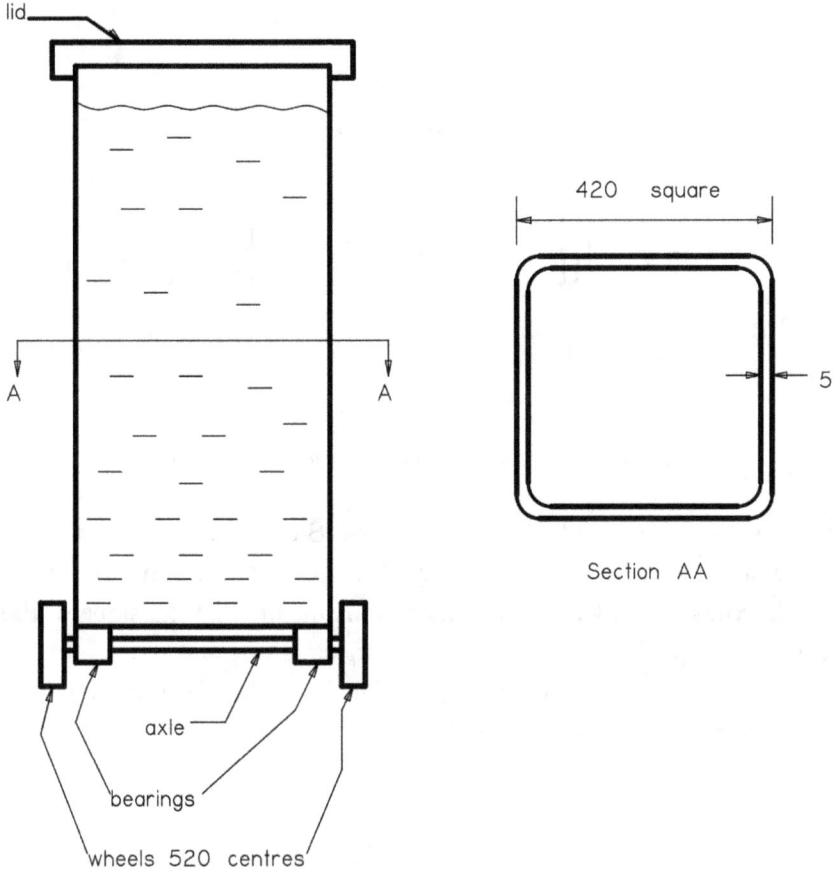

lid

420 square

5

Section AA

axle

bearings

wheels 520 centres

Figure 9.4.1 Wheelie bin full of water and its cross section.

Take a strip of the bin 1 mm deep near the bottom. The load on any side is reacted by tension in the adjacent sides; since adjacent strips will deflect by nearly the same, we ignore load transfer to them. Consider the side AB under the pressure p as shown in Figure 9.4.2 below.

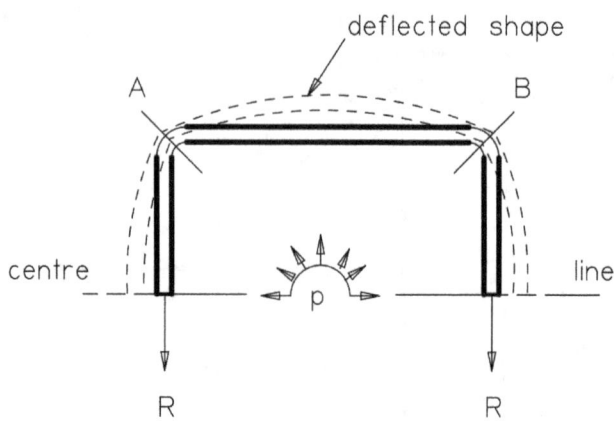

Figure 9.4.2 A 1mm deep strip near the bottom of the bin

AB is a beam which is fully fixed at A and B because of the symmetry of the structure and of the loading. Every side is the same in length and thickness and identically loaded. This means that the slope of the deflected strip and that of the adjacent ones are unchanged at A and B, which is the definition of fully fixed. The central deflection of a fully fixed beam under a uniform load is

$$defl = \frac{wl^4}{384EI}$$

l is the length, 420mm; *w* is the pressure x 1mm width of strip, and the pressure for 0.9 m of depth of water is 0.9 x the 1m value calculated above

$$w = 0.9 \times 0.009810 \times 1 = 8.829 \times 10^{-3} \text{N/mm}^2$$

E for polypropylene is 1300 N/mm².

$$I = \frac{width \times t^3}{12} = \frac{1}{12} \times 5^3 = 10.417 mm^4$$

$$defl = \frac{8.829 \times 10^{-3} \times 420^4}{384 \times 1300 \times 10.417} = 52.83 mm$$

This is ten times the thickness and so well outside the theoretical applicability of bending theory; however, it is large. If the deflection of the whole side is calculated as a complete plate, similar size numbers result, which confirms this one. The lengthening of the strip due to this transverse deflection, if it were to happen, can be calculated as approximately 17.50mm.

Next we compare this result with the deflection if the bin side were acting solely as a membrane. First we calculate the membrane stress. In Figure 9.4.2, the two Rs react the pressure on the strip

$$R = \frac{wl}{2} = 8.829 \times 10^{-3} \times \frac{420}{2} = 1.854N$$

The stress on the strip:

$$membrane\ stress = \frac{1.854}{5 \times 1} = 0.3708\ N/mm^2$$

We realise that all four sides are stressed identically, so the reactions R act on every side.

The deflection due to this purely tensile stress is

$$defl = \frac{\sigma r}{E}$$

σ is the stress, also known as hoop stress, and r is the radius of the circular shape which the strip takes up. This is a well-known result for cylinders under pressure.

$$hoop\ stress, \sigma = \frac{pr}{t}$$

Extracting r from that:

$$r = \frac{\sigma t}{p}$$

Substituting for r in the deflection equation gives the following:

$$defl = \frac{\sigma^2 t}{pE} = \frac{0.3708^2 \times 5}{0.00981 \times 1300} = 0.05391mm$$

This is a lengthening of the strip, and it is much smaller than the figure derived from the bending calculation (17.50mm), showing that this is a stiffer load path. In pure bending, there is not assumed to be any lengthening, but hypothetically the deflection calculated above of 52.83 mm can be put into the intercepting chord theorem, ac = bd of Figure 9.1.8, to find the radius of curvature, assuming this to be part of a circle, to find the length of the chord. Subtracting the original length of the strip gives 17.50mm. Try it.

Therefore, most of the load will be resisted by membrane action. However, because all the sides bulge out, it can be seen that there must be bending at A and B in Figure 9.4.2. We can estimate the bending stress at the corners by using the membrane deflection in the bending calculations.

It is easy to calculate the fixing BM for a fixed-ended beam such as our strip. We then reduce this in the ratio of the two lengthening deflections. We can do this because both are assumed to be based on the linear part of the stress/strain curve for the material.

$$fixed\ ended\ M = \frac{p \times l^2}{12} = \frac{8.829 \times 10^{-3}}{12} \times 420^2 = 129.8Nmm$$

The ratio of the strip lengthening is

$$ratio = \frac{0.05391}{17.50} = 0.003081$$

The moment is therefore

$$M = 129.8 \times 0.003081 = 0.3999Nmm$$

The usual bending stress equation gives us the following:

$$\sigma = \frac{My}{I} = \frac{0.3999 \times 2.5}{10.417} = 0.0960 \, N/mm^2$$

This will be tension on the inside of the corner. It must be added to the hoop stress, which is 0.3708N/mm². The total stress is 0.4668Mpa. This is unfactored. The Safety Factor is five, as there is no safety connotation. Comparing this with the minimum allowable ultimate stress of 21Mpa gives a RF of more than 10 – well in! This justifies all the approximations made in this study. I chose the minimum ultimate allowable, as there is no knowing where the bin was made or how it has been used.

This leaves the bottom, axle, wheels, and lifting rim to stress. If the bottom is spherically shaped, then it can be stressed just as the chair pan was. Find the radius by measuring the width and depth of the bottom below the level of the sides; use the intercepting chord theorem to calculate the radius and then the stress. Remember that this is half the hoop stress:

$$stress = \frac{pr}{2t}$$

p is the pressure, r is the radius, t is the thickness. Try it yourself.

The remaining items depend on the weight of the water in order to calculate the loads:

$$wt = volume \, \times density$$

$$wt = 0.42^2 \times 0.9 \times 1000 \, kg/m^3 = 159kg$$

Don't let it fall on you! This bin is smaller than standard, by the way; I've measured my own bin.

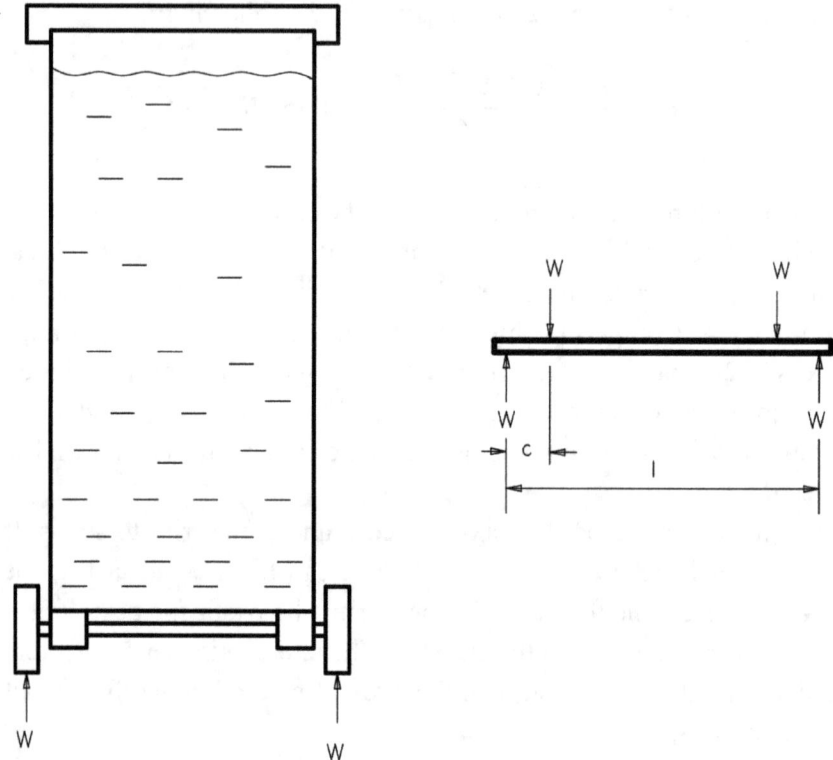

Figure 9.4.3 The bin on its wheels and the axle separately

The weight of the bin and water loads the axle through the bearings and c measures from the middle of the axle to the middle of the bearing. The BM obviously rises from the left end up to the position of the bearing, and then it stays constant until the right-hand bearing, when it declines to zero. The situation is symmetrical. The pair of loads at each end is called a couple, which is the same as a bending moment. The value of c is 85mm, W is half 159 kg, and so the couple is

$$\frac{159 \times 9.81 \times 85}{2} = 66291Nmm$$

The *I* of a tube is

$$\frac{\pi}{64} \times \left(D_o^4 - D_i^4\right) = \frac{\pi}{64}(20^4 - 16^4) = 4637mm^4$$

The outer diameter, D_o, is 20mm, and the inner, D_i, is 16mm. This means the tube thickness is 2mm. The bending stress all along the tube between the bearings is as follows:

$$stress = \frac{My}{I} = \frac{66291 \times \frac{20}{2}}{4637} = 143\ N/mm^2\ or\ MPa$$

The Safety Factors must be chosen next. Depending on the use intended for the bin, safety may or may not be an issue. If it is and considering the unknown manufacturing history, it is prudent to choose an SF of 3 ultimate and 2 yield. Since the material standard is probably unknown, we must assume mild steel with an ultimate allowable of 465Mpa and a yield of 230Mpa (see table in Chapter 8).

Ultimate: stress is 429Mpa, allowable 465Mpa, and the RF of 1.08 is fine.

Yield: stress is 286Mpa, allowable 230MPa, and RF of 0.80 is not good.

Fatigue: if there is a case, stress is 143Mpa and allowable 233Mpa. This is okay, but the usage of the bin is important. If it is pulled over rough ground, the stress will be higher. Also, the axle needs to be non-rotating; otherwise, there is a fully reversed stress cycle all along it. The combination could cause fatigue failure.

In order to solve the yield problem, the easiest thing to do is change to a solid bar. The *I* is as follows:

$$I = \frac{\pi}{64}D_o^4 = \frac{\pi}{64}20^4 = 7854mm^4$$

The stress is reduced inversely as the *I*s are increased.

$$stress = 143 \times \frac{4637}{7854} = 84\,N/mm^2 \ unfactored$$

This is 168N/mm² factored and gives an RF of 1.37. The other RFs are improved also.

The wheels could be a problem. It would probably be best to buy new wheels to go with the new solid bar axle. You would choose ones that had the load capacity to support half of 159 kg (80 kg minimum); a search on the Web for trolley wheels should turn up something suitable.

The rim at the front top of the bin, which is normally used to lift it and tip its contents into the lorry, will also need analysis if the bin is required to be lifted.

Of course you could just fill the thing with water and see what gives!

5. The spoked wheel

Ch 1   Load cases

Consider a spoked wheel. There are two sorts; the first is like an old-fashioned cart or a modern car. It has integrated spokes and rim. The second has slender spokes like a bicycle. The load is between the road and the hub via the spokes. The load in a spoke varies, as the wheel spins and bumps in the road need to be taken into account. Braking and acceleration loads will have to be considered.

Ch 2   SF

Most modern spoked wheels are made from metal, mostly steel or aluminium alloys. The choice of yield and ultimate factors from the table depends, therefore, on the method of manufacture. Safety must be a consideration. Fatigue is bound to be important, and the stress should be less than the fatigue limit, which should be used with a factor of 2; this will probably be the critical case.

Ch 3   Structural data

The shape of wheels with integrated spokes and rim, as seen on fast or fancy cars, can be quite complicated. Drawing and measuring them from an example would be a lengthy business; if you have manufacturing drawings, so much the better. The exact grade of the steel or aluminium will have to be established.

Wire spoked wheels are simple to measure, but again, the material grade of the spokes is important; it will be of a high strength, I expect. The literature should be consulted. The number of spokes will be needed, as will their inclination to the radial direction.

Ch 4   Classification by component type

If the spokes are wire (that is, slender, as in a bicycle wheel), they will be in tension both from the initial pre-tension and the weight of bike and rider (this interaction is discussed in Chapter (6) Section (g)). The rim will be in compression and bending. If the spokes are cast integrally with the rim and are relatively wide, there will be bending as well as tension and compression. The rims are curved beams in the gaps between the spokes.

Ch 5   Support reactions

The reaction to the axle load of a wheel is, obviously, at the road. If the wheel is hard like a steel train wheel, the reactive load will approach a point load as far as simple stressing purposes go. If the wheel is soft like a pneumatic tyre, the load will be spread over a patch of road and rim whose size depends on the pressure in the tyre and its wall stiffness.

Ch 6   Loads and their application

Assuming the wheel has a tyre, the rim will be loaded by a distributed pressure load and a frictional interaction between the tyre walls and the rim walls. This is too complicated for this exercise – it needs computers. For this example, we will depend on test results for the support assumptions. The other important component is the spoke. This will experience varying levels of tension.

Ch 7   Calculations

For this example, let us take a spoked bicycle wheel and assume we want to use it for some purpose other than as a bicycle. The question is, how much weight can we put on it? For normal use, this is going to be about 50 kg maximum – that is, basically, a man balanced on both wheels. In fact, the actual maximum load is going to be much higher, although to an unknown level. He may lean over the front wheel and hit a bump; that could be more than twice the 50 kg (100 kg+). Say we need more than that. For simplicity, we will take the case of a radially spoked wheel.

The method which is used to transfer the weight of the rider from the hub to the ground is similar to that in a pre-tensioned bolt. Each spoke is tensioned up sufficiently not to become slack when it is at the bottom next to the ground. Actually, it should be tightened as much as possible to reduce the fatigue damage in the rim and spoke; this is the same as and was explained in the section on pre-tensioned bolts (Chapter 6 section (g)). Also, of course, the spokes have to hold the rim truly round and in one plane, but that is not our concern here. The rim is compressed by all the spokes, which act much like a pressure on the inside of a cylinder. This is shown in Figure 9.5.1 (a). Here the rim deflects inwards to the dashed line and the spoke outwards to meet the deflected rim.

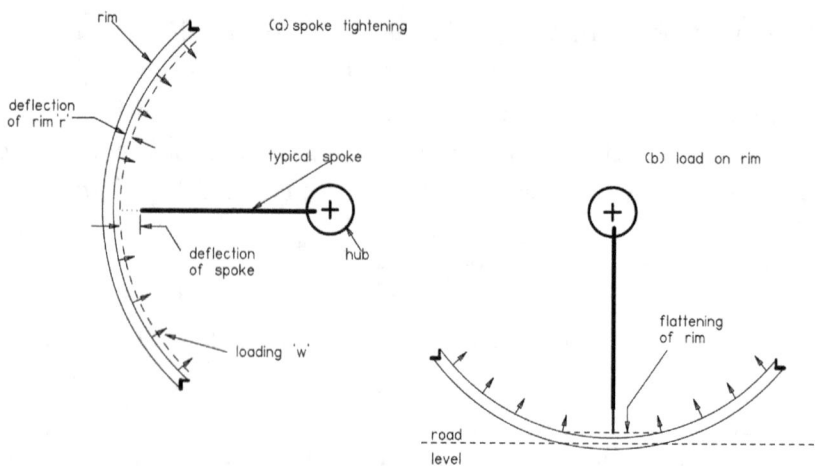

Figure 9.5.1 Part of a spoked bicycle wheel: a) pre-tensioning of the spokes, b) with load applied, showing the (exaggerated) flattening of the rim

When the load is applied at the road, consider the situation when the wheel is stationary. The rim will bend a little so that the distance from rim to hub of the spoke right at the bottom is shortened, and therefore the spoke loses some of its tension. Surrounding spokes may also be somewhat shortened – see, Figure 9.5.1 (b).

This situation has aroused some interest over the years, as it was not obvious how the wheel worked in its job of carrying a load. It was in the second half of the nineteenth century that the pre-tensioned spoked wheel was developed; however, I guess it was done by trial and error, not calculation. Tests and simple FE calculations in the following century, available on the WWW if you're that interested in the detail, show that the bottommost spoke is relaxed from its pre-tension the most, the spoke on either side less so, and all the others have their pre-tension increased. This is to be expected, as the bending straightens the rim. Obviously, the straightened piece of rim is shorter than the arc it described before. This difference in length, an increase, will increase the diameter (as the total length of rim will remain nearly the same), and thus the other spokes will be further extended.

The true situation would take a computer programme, a Finite Element calculation, and a non-linear one at that, to solve. This is because it is not obvious how much of the rim will bend and how the tension in all the other spokes changes; there are also other matters to take into account: the tyre and its pressure, the transverse flexibility of the rim, the flexibilities of the hub and fixings. The non-linear FE finds out by a sort of mathematical trial and error method. It's very complicated.

Much simpler and therefore more approximate methods follow. We will consider three methods.

Firstly, get a spoke tension measuring tool (for instance, a Parktool TM1) and measure the tension in the spokes to verify the value and that all spokes are reasonably equally tensioned. This device bends a section of the spoke between three points and returns a number. Look up the number on a supplied chart and you get the tension in kg. The spoke diameter and material (steel, titanium, or aluminium) are also factors; there is a separate chart for each combination. The tensions usually vary between 80 kg and 230 kg. The theory behind the method is that of a beam under a point load and in tension. The tension pulls out the bending deflection – that is, it reduces the deflection to a calculable extent. This has been done for the charts.

Now, most of the theoretical and practical tests show that the wheel load to slacken the spoke is at least 1.75 times the pre-tension load. Therefore, it seems safe to specify that the wheel loading could be one and a half times the pre-tension as a limit load. The spoke and rim must then be stressed to show that the new higher wheel load that you want to apply is safe.

Secondly, if you do not have access to a tension meter, here is a practical method, take your wheel and slacken a spoke until it is just about to be loose, counting the number of turns of the spoke's nipple. You will first have to mark the nipple and rim so that you can return to the original position. It is also a good idea to ensure that all the spokes have the same pre-tension. Bicycle spokes all seem to have the same thread (a UNC-2B unified thread), with 56 tpi (fifty-six teeth per inch). This is 2.206 teeth per millimetre or 0.4534 mm per complete turn. When all the nipples are just undone, the spokes lengthen and the rim increases its diameter. The sum of these two deflections equals the tightening of the nipple. We will construct this equation, which will yield the spoke load.

Figure 9.5.1(a) shows the two deflections of interest. The deflection of the rim radially under a hoop compression is (from standard equations found in *Roark* and other sources) as follows:

$$rim\ defl = \frac{\sigma_h}{E_r} R$$

$$Rim\ stress, \sigma_h = \frac{w \times R}{A_r}$$

Now we substitute the value of σ_h from the second of these equations in the first, giving us:

$$rim\ defl = \frac{wR^2}{A_r E_r}$$

In these, σ_h is the stress in the rim due to the tightening of the spokes. R is the radius of the rim from the wheel centre to the centroid of the rim cross section; w is the (assumed continuous and uniform) loading on the rim from the spokes' pre-tension and has units of N/mm of circumference.

E_r is the Young's Modulus and A_r the cross-sectional area of the rim.

If the number of spokes is n, then the distance between spokes is the circumference divided by the number of spokes:

$$spoke\ pitch = \frac{2\pi R}{n}$$

If the pre-tension load in the spoke, call it P, is the loading, w, times half the length of the rim each side of the spoke, which is the same as the pitch.

$$P = \frac{2\pi R}{n} \times w$$

The extension of a spoke:

$$spoke\ extension = \frac{Pl}{A_s E_s}$$

Here, l is the length of the spoke from its anchorage in the hub to the nipple when this is just finger tight. The subscript s in the A and E terms describes the spoke. Substituting for P in this equation gives us

$$spoke\ extension = \frac{2\pi R w l}{n A_s E_s}$$

The sum of the two deflections, rim and spoke, equals the measured deflection – that is, the number of complete turns of the nipple times 0.4534 mm, as described above. Call that T:

$$\frac{wR^2}{A_r E_r} + \frac{2\pi Rwl}{nA_s E_s} = T$$

And with minimal manipulation, extracting w from the expressions on the LHS gives us the following:

$$w = \frac{T}{\dfrac{R^2}{A_r E_r} + \dfrac{2\pi Rl}{nA_s E_s}} N/mm$$

The w can then be substituted back into the equation above for P.

The way to tackle the arithmetic is to assemble the data neatly and substitute into each of the separate rather lengthy expressions. I have taken the liberty of using the wheel data from a paper by Burgoyne and Dilmaghanian (*Journal of Engineering Mechanics*, vol. 119, no. 3, March 1993), who performed a series of tests and analyses on a bicycle wheel.

$R = 309.4$mm; $A_r = 138.4$mm²; $E_r = 70,000$N/mm²; spoke diameter = 2.1mm so
$A_s = 3.464$mm²; $E_s = 210,000$N/mm²; $l = 286.5$mm; also take n=36:

$$\frac{R^2}{A_r E_r} = \frac{309.4^2}{138.4 \times 70000} = 0.009881$$

$$\frac{2\pi Rl}{nA_s E_s} = \frac{2\pi \times 309.4 \times 286.5}{36 \times 3.464 \times 210000} = 0.02127$$

If the spoke is stretched by, say, 0.4534mm, exactly one turn of the spoke nipple, then:

$$w = \frac{0.4534}{0.009881 + 0.02127} = 14.55 N/mm$$

The spoke pre-tension load is now (from the equation developed above)

$$P = \frac{2\pi R}{n} \times w = \frac{2\pi \times 309.4}{36} \times 14.55 = 785.7 N$$

This suggests that the limit load is one and a half times as big (1179N), as per the tests, and is subject to the usual Safety Factors. Conventionally kg are used, so this translates to 118 kg limit load.

Thirdly, and for more accuracy, we can use the same method as for a pre-tensioned bolt and construct a load-deflection diagram for the spokes and rim. We have the deflections for the pre-tensioning now that we have the value of w:

$$rim\ defl = \frac{wR^2}{A_r E_r} = \frac{14.55 \times 309.4^2}{138.4 \times 70000} = 0.1438 mm$$

$$spoke\ extension = \frac{2\pi Rwl}{nA_s E_s} = \frac{2\pi \times 309.4 \times 14.55 \times 286.5}{36 \times 3.464 \times 210000} = 0.3094 mm$$

These are for a pre-tension load of 785.7 N and for one turn of the nipple. We can plot these deflections on a graph against load, as in Figure 9.5.2.

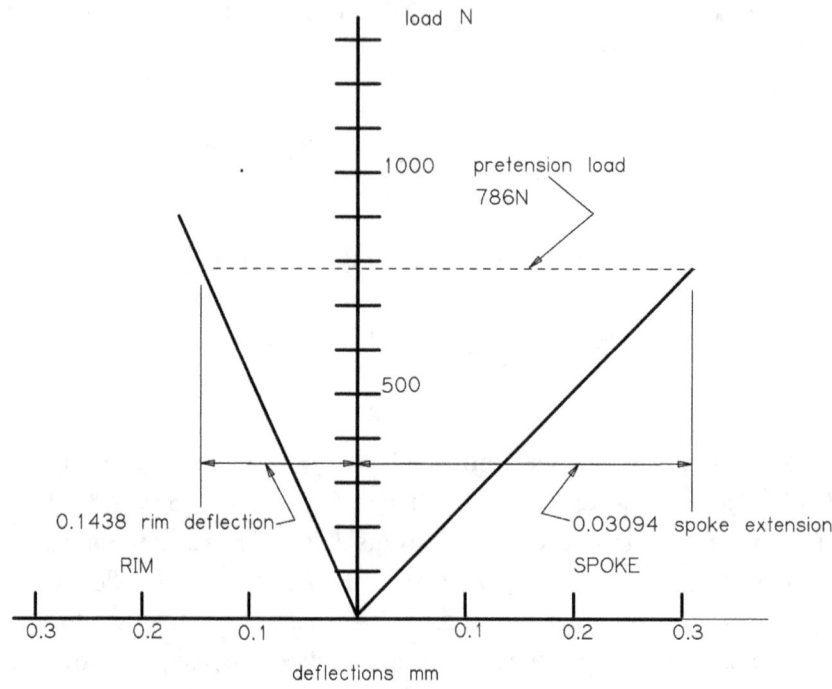

Figure 9.5.2 Plot of pre-tension deflections of rim and spoke for the 786 N load; the spoke's tension is to the right and the rim's hoop compression to the left.

Obviously, the deflections are zero at zero load, and as the load increases, the deflections increase linearly up to the 786 N line and beyond until something yields or breaks. This is the situation with no load applied to the bicycle.

When a load is applied, as has been previously stated, the rim bends a little around the bottom spoke. This is an additional deflection. The next job is to calculate this bending deflection; we will do this at the same load as the pre-tension load for convenience – we can then more easily add it onto the previous graph (Figure 9.5.2).

All the tests and computer calculations show that most of the weight of the rider is reacted by the bottom spoke, most of the other spokes only experiencing a small uniform increase in tension. We shall assume that the entire load is taken by the bottom-most spoke. Actually, the load is a reduction in tension; the load is taken by the difference in tension

270

of the spoke. Furthermore, we shall assume that this is equivalent to a point load applied to the rim. In practice, the rim will experience a distributed load, as the inflated tyre (itself a pre-tensioned system) will spread the load, and the fact that the rim bends and straightens means that the load is also spread a little. But these assumptions will ensure a conservative situation because a point load will give a larger deflection than a distributed load. So we will take the rim as a beam under a point load and fixed against bending at its ends. Since the rim depth in the radial direction is a small fraction of the wheel diameter, it is also acceptable to assume that the beam is straight. To allow for the fact that the two spokes adjacent to the bottom spoke take some load, the length of the beam is taken to be three spoke pitches.

The equation for the deflection of such a fully fixed beam is

$$\delta = \frac{Pl^3}{192 E_r I_r}$$

P is the load, E_r is the Young's modulus, I_r is the second moment of area (it is 1469mm⁴), and l is three times the distance between two spokes at the rim centroid.

$$l = 3 \times \frac{2\pi R}{n}$$

$$l = 3 \times \frac{2\pi \times 309.4}{36} = 162.0$$

So the bending deflection is

$$\delta = \frac{786 \times 162^3}{192 \times 70000 \times 1469} = 0.1693mm$$

We can plot this on the previous graph to give Figure 9.5.3. This is the line from the origin through 0.1693 at e and extended on. It is labelled as 'bending'.

Notice that had we assumed that the loading on the rim was partially distributed the deflection line would have been steeper – that is, a given deflection would need a larger load.

But in accordance with the test evidence, we will assume that any applied load only bends the rim. This follows the line in the Figure labelled bending. The rim and the other spokes are assumed to be unaffected and to stay in their tensioned state. So any deflection of the rim generates a load which can be read off on the vertical axis.

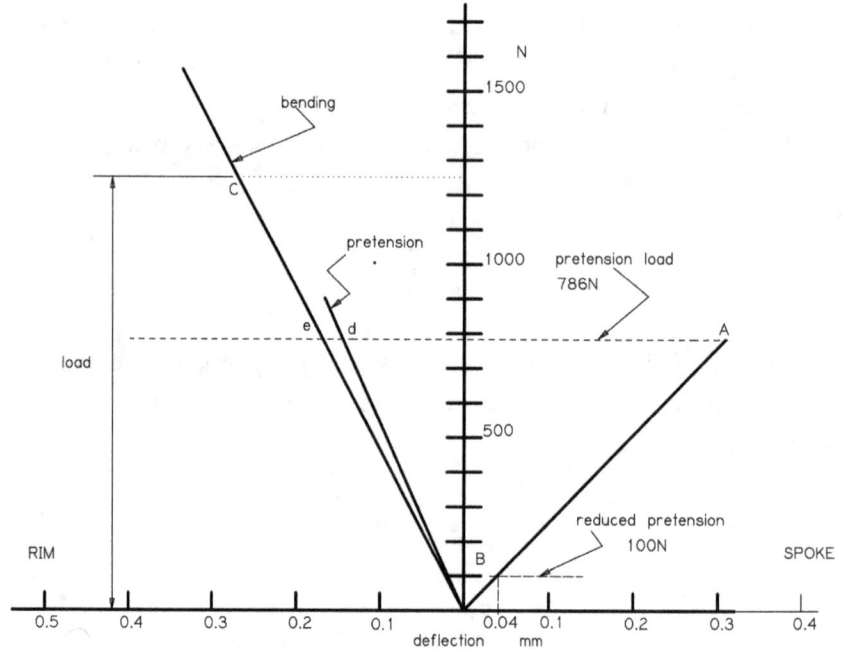

Figure 9.5.3 Finding the allowable load on the wheel graphically

The next step is to figure out the load that can be applied to the wheel without slackening the spokes completely. As an example, we will reduce the load in the spoke to 100N, and from the spoke's deflection line, we see that the deflection remaining is 0.04mm. The initial deflection was 0.3094mm, so the change is 0.3094-0.04 = 0.2694mm. This change is, according to our assumption, also the change in deflection of the rim. Point C marks the point on the bending, line which is 0.2694 (nearly

0.27), which gives a load of 1250 N – just a little more than the previous estimate.

To sum up, a pre-tension of one turn of the spokes' nipples allows a limit load of 125 kg; this is an arbitrary choice of pre-tension, of course, so the next thing to do is check some stresses and see whether it ought to be changed.

We will find the stress in two components: the tension stress in the spokes and the bending stress in the rim. For this, we need to assemble the required data.

The spoke stress requires the cross-sectional area and the load. Both are given above.

$$spoke\ stress, \sigma_s = \frac{P}{A_s} = \frac{786}{3.464} = 227\ ^N/_{mm^2}$$

The material of the spoke is probably a drawn (as in pulled) stainless steel wire. This means its strength is enhanced because the drawing, a method of manufacturing, work hardens the steel. For a typical hard drawn wire of 2 mm or so diameter, strengths could be: ultimate strength 2050N/mm² and proof strength of 1740N/mm². The fatigue limit is 860N/mm². If a spoke breaks, the situation is not too serious, and they are precision engineered so the Safety Factors need not be too high. We will take 1.5 for the ultimate case and 1.25 for the proof. For the fatigue case, take a SF of 2.0. If this is too limiting, a more detailed analysis will have to be undertaken.

The Reserve Factors are the allowable stresses divided by the actual stresses times their Safety Factors.

$$Ultimate\ RF = \frac{2050}{227 \times 1.5} = 6.02$$

$$Proof\ RF = \frac{1740}{227 \times 1.25} = 6.13$$

$$Fatigue\ limit\ RF = \frac{860}{227 \times 2} = 1.89$$

The critical RF is obviously the fatigue one, which means the maximum load is 1.89 times 786N, or 1486N, considering only the spoke stress. The others would give more than three times that: 4730N and 4820N. Since we are being conservative, it is likely that doing a detailed fatigue analysis would give a higher result. But first it would be sensible to look at the rim bending, as this may give a lower load.

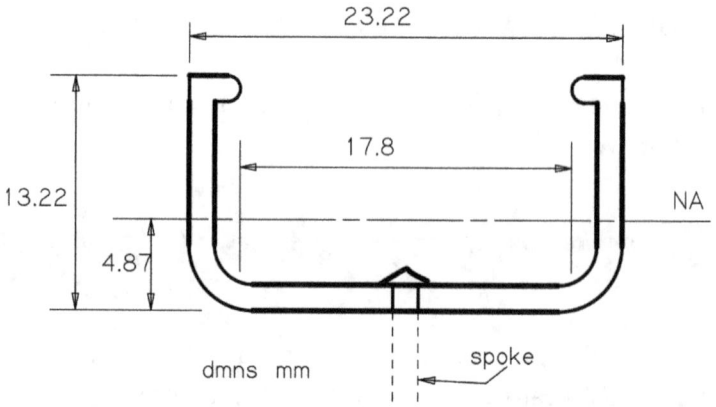

Figure 9.5.4 Cross section of bicycle rim – not to scale

The rim bends towards the spoke so the horizontal flat, which connects to the spoke, is in tension. This is important because the flat contains the hole which becomes a stress concentration, which must be considered. There are two stress systems to consider; the first is the bending, and the second is the initial hoop compression.

The bending moment for a fixed-ended beam with a central point load is

$$M = \frac{1}{8}Pl = \frac{1}{8} \times 786 \times 162 = 15917 Nmm$$

Bending stress on the tension side is

$$bending\ stress = \frac{My}{I} = \frac{15917 \times 4.87}{1469} = 53\ ^N/_{mm^2}$$

Here y is the distance from the neutral axis to the lower face, and I (1460 mm^4) is the second moment of area of the cross section, as usual. The hoop stress is compression and will subtract from this tension:

$$\sigma_h = \frac{wR}{A_r} = \frac{14.55 \times 309.4}{138.4} = -33 \, {}^N\!/_{mm^2}$$

The net stress is only 20N/mm^2. But there is the stress concentration factor, which is given in many publications. If the rim were infinitely wide, it would be at its maximum of 3.0. The formula in *Roark* based on tests is

$$k = \frac{3d}{a + d}$$

k is the factor, d is the width of the plate, and a is the diameter of the hole. You can see that this tends to 3.0 as d increases to a large value:

$$k = \frac{3 \times 17.8}{2.2 + 17.8} = 2.67$$

The spoke diameter is 2.1mm, so I have assumed the hole is a little bigger. The maximum stress is

$$stress = 20 \times 2.67 = 53 \, {}^N\!/_{mm^2}$$

It is a coincidence that this is the same value as the bending stress. Compare this to the fatigue allowable, as this stress at a hole occurs once for every turn of the wheel.

A likely aluminium alloy has strength properties of 310N/mm^2 ultimate strength, 260N/mm^2 proof and 186N/mm^2 limit fatigue. We can use the same SFs as for the spoke:

$$Ultimate \; RF = \frac{310}{20 \times 1.5} = 10.3$$

$$Proof\ RF = \frac{260}{1.25 \times 20} = 10.4$$

$$Fatigue\ RF = \frac{186}{53 \times 2} = 1.75$$

The other side of the rim is, of course, in bending compression as well as hoop compression:

$$\sigma_b = \frac{My}{I} = \frac{15917 \times (13.22 - 4.87)}{1469} = 91\ ^N/_{mm^2}$$

The hoop compression is the same, so the total stress is $91+33=124N/mm^2$:

$$Ultimate\ compression\ RF = \frac{310}{124 \times 1.5} = 1.67$$

$$Proof\ compression\ RF = \frac{260}{124 \times 1.25} = 1.67$$

These are the lowest Reserve Factors and that means the largest load we should put on the wheel is 1.67 x 786 = 1313 N or 131 kg.

If the spokes are not radial (most aren't), as a first approximation, the load should be divided by the cosine of the angle at the rim between the radial spoke and actual spoke. This also applies to the angle between the plane of the rim and the spoke. These effects are usually small, a few degrees each, and make only a per cent or so difference.

Summary

These examples have been worked through in detail in order to demonstrate the various techniques which were introduced in the book. The detail would probably be thought excessive by a professional stressman, but the book is not, as stated at the beginning, for professionals, although they may use these sorts of calculations to get a preliminary understanding of a new structure in new situations.

Acceleration The rate of change of velocity – that is, the amount velocity changes in a time period. Units are metres per seconds squared, or m/s^2.

Alternating stress Part of the total stress used in a fatigue assessment. It is the stress that does most of the fatigue damage. It is half of the difference between the maximum and minimum stresses in a fatigue cycle. This stress alternates about the mean stress: alternating stress = ½ (max stress – min stress)

Amplitude The peak value of a varying quantity; often used in dynamics.

Balance equation This equates all the forces in one direction with all those in the opposite direction; it is Newton's Third Law.

Bar Carries tensile or compressive loads.

Beam Carries loads which bend it as well as possibly end loads (tensile or compressive).

Bending When structures or components try to curl up. The stresses on one side of the Neutral Axis are tensile and on the other side compressive.

Bending Moment (BM) The tendency for a load to cause bending. A bending moment is a load multiplied by a length. The load can be of any type, UDL, point, and so forth. See also torsion and torque. Units are Newton x metres, or Nm.

Bolt hole fit Some bolts do not necessarily touch the sides of a hole; they transfer their shear load by friction under their heads. Others are in holes which have been carefully lined up and sized to fit as exactly as possible, even by forcing the bolt into the holes. There are recognized in-between situations. For instance, these are detailed in *Machinery's Handbook*.

Brittle Refers to materials of every sort which fracture without much distortion of the failure region.

Buckling Can take place when a component or structure is in compression. The stress may be less than yield.

Cantilever A beam, plate, or structure which is supported at one end only. There is a shear support and a moment support.

Centroidal axis The centre refers to a cross section of a bar or beam. In a bar, it means the centre of area; this is not necessarily the same as the neutral axis in a beam.

Component This has two meanings according to context. Firstly, it may be part of a bigger structure, anything from a simple beam to a complex structure itself. Secondly, it is part of a load which has been split into two or more directions.

Composite materials More than just metal or plastic by themselves, a composite is made up of at least two different materials. The most common composite is probably glass fibre in a plastic matrix, but the combinations are endless.

Compression Shortens structures.

Constraints The supports to a structure. They must carry the loads on a structure, both direct and rotational, so that Newtons Third Law is satisfied.

Conversion factors Convert quantities from one system of measurement to another. There are many, the most common being the imperial and the metric systems. A table is given in Chapter 7, section b.

Couple Consists of two equal loads separated by a distance. It is calculated by multiplying them together, so its units are Nm; it produces bending when applied to, for instance, a beam.

Creep A permanent deformation in a structure under load. It takes time and an elevated temperature. Unreinforced plastics creep at room temperature, but metals creep at higher temperatures.

Cross section This is produced by visualizing a cut face across a component such as a beam. It yields information needed to calculate the stress in the component (for instance, its area, second moment, and position of the neutral axis).

Damping The action which kills vibrations. It may be slight, which means the vibrations carry on for a long time, or so strong that only half a cycle is completed.

Degree of freedom In this subject, it refers to the components of movement or deformation of a structure. There are six degrees, three linear along axes perpendicular to each other and three rotations about each of these axes.

Density The mass per unit volume of a substance in any phase, gaseous, liquid, or solid. Units are kg/m^3.

Determinate A structure when it can be analysed by the balance equations alone.

Ductile Materials stretch considerably before fracture. The non-elastic region is large compared to the elastic in metals.

Edge conditions A description of how a structure is supported at its edges. See constraints (above). They include rotational as well as direct supports.

Elastic limit In materials, refers to the point on the stress/strain curve at which the material ceases to behave elastically – that is, if the stress is removed, the structure does not return to its original shape. In steel, there is usually a marked change of behaviour, but in other alloys, the transition occurs gradually, so a point has to be chosen which is deemed to be that limit. This is when the permanent strain is 0.1 per cent or similarly small.

Elements Part of an FE model. They define the material properties and sizes of the structure under analysis.

Fatigue A failure condition suffered by most materials. It is due to a repeated loading, and the stress at failure is less than the yield or ultimate failure stresses.

Finite element computer program (FE) a method of analysing complicated structures in which the structure is represented by a series of points (nodes) connected by the elements. It uses so-called numerical mathematics and needs computers to solve the numerous simultaneous equations which are generated by the program.

Fixed The edge condition or constraint that applies a moment restraint to a component such as a beam or plate. The size of the moment is such as to allow no rotation of the beam there.

Fluid A gas or a liquid.

Hertzian stress Stresses at a point contact, outside the scope this book.

Hoop stress In a curved plate or beam caused by a pressure or similar loading (see example of bicycle wheel). It is in the circumferential (or meridional) direction. In a circular tube, hoop stress equals pressure times radius of tube divided by tube thickness.

Units are the usual for stress, N/mm^2. In a sphere, the hoop stress is half that.

I A property of the cross section of a beam used to calculate the bending stress. It is strictly the second moment of area of the cross section and colloquially known as its inertia.

Impact A load which occurs for a very short time.

Indeterminate Structures or components which cannot be analysed by the use of the balance equations alone. They are solved by using deflections to produce more equations. This book only gives the results of these solutions.

Limit of proportionality Another way of describing the elastic limit (see above).

Line diagrams of structures. When structures and single components are drawn, the only lines of interest are the centroids of beams and bars and the mid-plane of plates. All other details are irrelevant clutter. The only exceptions are compact items such as lugs; even with these, only the outlines are drawn.

Line loads Figure 6.1 shows various ways these can be represented. This can be extended into the third dimension for plate loadings.

Linearity In equations, this refers to the ratio between two quantities when one is directly proportional to the other; if one doubles, so does the other. Also, if a structure behaves linearly, then the stresses and strains due to two or more loads can be added together.

Manufacturing tolerance Nothing can be made exactly, so all dimensions have an upper and lower limit. It is sometimes necessary to use the lowest thickness of a plate in stress calculations when sizing it.

Mass The quantity of stuff in any item. Units are kg.

Mean stress In a fatigue case, this has to be extracted from the fatigue cycle, as it is the alternating stress which does most of the damage. Mean stress = ½(max stress + min stress).

Meridional stress Same as hoop stress.

Mesh To mesh the drawing of a structure is to choose all the nodes and join them together, thus forming the elements. Nowadays meshing can nearly but not quite adequately be done automatically by a computer program.

Mode In a vibrating structure, this is the shape of the distortion.

Natural frequency The number of cycles of deflection per second that a vibrating structure displays when it is free of any other force or influence. You bang the drum and lift the stick away and the drum vibrates at its natural frequency, which depends on its size and construction. Units are cycles per second, or Hertz.

Neutral axis The point on the cross section of a bending beam which has zero stress. On one side, the stress is tension, and on the opposite, it is compression.

Newton In SI units, it is the unit of force. Using Newton's Second Law it is a kilogram times metre/second^2. See table Chapter 7, section (b).

Nodes The representative points on a structure. They define its geometry.

Orthogonal Axes are at right angles to each other.

Partial structure In a complicated structure, individual components or sets of components can be extracted and analysed if the edge conditions and loads from the rest of the structure are applied.

Plastics A non-metallic material, usually organic carbon based and consisting of long chained molecules twisted around each other, known as polymers. They can be used to make structures by themselves or reinforced by fibres of other materials. Glass fibre reinforced polyester is probably the most usual combination, but I expect all possible combinations have been tried.

Point load A force which is considered to act on a structure at a point. Away from the point, this is perfectly acceptable; at the point, it is not, as it implies infinite stress. See Hertzian stress.

Poisson's ratio The ratio of the transverse deformation to the longitudinal deformation under a longitudinal stress.

Preloads Loads built into a structure, usually during manufacture.

Pressure Any load spread over an area. Units are N/m^2.

Principal stress When there are forces from two or more directions on a body, it is not clear what the maximum stress is. There are equations involving the angle of each force as well as its size which give this value. Also given are the minimum stress, the shear stress, and all their angles.

Proof condition For materials which do not exhibit a marked transition from linear stress/strain behaviour to the non-linear, where permanent set occurs, an arbitrary strain limit is set, such as 0.1 per cent strain. The calculated stress in any load case is compared to this. See elastic limit (above).

Propped cantilever As in the cantilever (above), but the free end is supported and not restrained rotationally.

Radius of gyration Technical term important in buckling considerations. Radius of gyration = square root of (least I of a cross section divided by its area).

Reinforced plastics See plastics (above). Reinforcing fibres are glass, carbon, aramids, boron, and so on.

Relaxation Inverse to creep. When some materials are under stress but are held immoveable, that stress reduces; it is time and temperature dependent.

Reserve Factor This is the allowable load or stress divided by the applied stress (which has been multiplied by the Safety Factor – hence known as the factored stress). If over 1.0, the structure is strong enough by the amount it is greater than 1.0. If less than 1.0, it is not.

Residual stress Leftover stress after loads, temperatures, or manufacturing processes are removed.

Right-Hand Rule An arbitrary way of arranging the directions (x, y, z) of the three Cartesian orthogonal axes. Spread the thumb, first, and second fingers at right angles to each other: x is the thumb, y the first finger, and z the second finger.

RSJ (rolled steel joist) A standard I-section beam much used in the building industry.

Safety Factor (SF) Applied to the loads on a structure during calculations. They increase the structure's strength to cover unknowns in all aspects of its life.

Second moment of area See I (above).

Shear A parallel sliding action or stress, at right angles to a tensile direction in a beam.

Simply supported (SS) Refers to a structure – say, a beam – which is supported without any rotational restraint or clamping action at its ends.

Slenderness ratio Indicates the susceptibility of a bar or plate to buckling. Slenderness ratio = length divided by least radius of gyration

Speed The rate of change of position – that is, the amount position changes in a given time period. The direction is not involved. It is what your speedometer in your car says.

Stiffness The force required to cause a deflection of unit length. Units are N/m.

Strain The deflection of a bar in tension or compression divided by the original length. It is connected to stress by Young's Modulus.

Stress Force per unit area. Units are N/m^2.

Stress concentration Any sudden change in the dimensions of a component is liable to cause a stress concentration, which is a local, sharp increase in stress.

Symbols Used in this book in equations; each stands for a number. Usually they come from the standard alphabet, but some widely used ones come from the Greek. These are σ (sigma) for direct stress, τ (tau) for shear stress, and ϵ for strain.

Symmetry If a structure is symmetric, then one part is a mirror image of another. A calculation can sometimes be simplified by substituting half of the structure with a few constraints.

Tension This lengthens structures.

Torque This twists structures.

Torsion A shear stress. Instead of faces trying to slide over each other linearly, they attempt to slide rotationally.

Ultimate condition When a component or structure finally breaks.

Uniformly Distributed Load (UDL) The same as line loads above.

Units Everything is measured in some sort of unit. The SI system uses metres, kilograms, seconds, and temperature as its basic units.

Velocity The rate of change of position in a given direction – that is, the amount position changes in a given time period. It takes into account both speed and direction.

Vibration A repeated movement or deflections of a structure. It can be regular in time, as in the minute hand of a clock passing, say, 12 o'clock, or irregular, as the bouncing of a car's wheel going over a bumpy road. Units are cycles per second, or Hertz.

Weight The mass of an item multiplied by the acceleration due to gravity. Interestingly, as an object goes higher and higher into space, its weight diminishes although its mass does not.

Yield condition Mid- and low-strength steels show a marked transition from linear and reversible stress strain behaviour to the non-linear phase. This is called yield, and its stress is the yield stress used as an allowable stress. Other alloys and very high strength steels use a proof condition (see above). Loosely the two terms are used interchangeably.

Young's Modulus (E) The ratio of stress divided by strain.

BIBLIOGRAPHY

Blevins, Robert D. Formulas for natural frequency and mode shape, Krieger Publishing Company, Florida, 2001.
ISBN 1-57524-184-6

Dieter, George E. Mechanical Metallurgy, second edition, Mcgraw Hill London, 1982.
ISBN 0-07-085158-1

Eckold, Geoff. Design and Manufacture of Composite Structures, Woodhead Publishing Limited, Cambridge, 1994.
ISBN 1-85573-051-0

Frost, Marsh and Pook, Metal Fatigue, Clarendon Press, Oxford, 1974.
ISBN 0 19 856114 8

BISPA, Iron and Steel Specifications, London, 1994
(or similar in other jurisdictions) (8th edition – old)
ISBN 0-9523872-0-4

Johnson, W. Impact Strength of materials, Edward Arnold, London, 1972.
ISBN 0 7131 3266 3

Machinery's Handbook, Industrial Press Inc. New York, (24th edition 1992).
ISBN 0-8311-2424-5

Peterson, R. E. Stress Concentration Factors, John Wiley & Sons, New York, 1901.
ISBN 0-471-68329-9

ALFED, Properties of Aluminium and Its Alloys (British), 2009 and **ASTM** (US)

Roark Formulas for stress and strain. Now in its eighth edition by Warren C. Young and Richard G. Budynas, it was first published in 1938. The fourth edition is the simplest for the purposes of this book and may be available second-hand.
Library of Congress catalogue card number 64-66027

G. Saville et al. "Safety in Pressure Testing," Seminar S503, IMechE

Timoshenko, S. Strength of Materials Parts 1 and 2, Van Nostrand Rheinhold Company, New York, 1980 and 1981, ISBN 0-442-08539-7 and 0-442-08540-0

Timoshenko, S., and **Woinowsky-Krieger, S.** Theory of Plates and Shells, Mcgraw-Hill, New York, 1959, Library of Congress catalogue card number 58-59675

Timoshenko, S., and **Gere, J.** Theory of Elastic Stability, Mcgraw-Hill, New York,1951,
Library of Congress catalogue card number 59-8568

Timoshenko S., Young D. H., and **Weaver W., Jr.** Vibration Problems in Engineering,
John Wiley & Sons, New York, 1974, ISBN 0-471-87315-2

"Bicycle Wheel As a Prestressed Structure." C. J. Burgoyne and R. Dilmaghanian, *Journal of Engineering Mechanics,* vol.119, no. 3, March 1993, paper No. 2425.

The Internet is a good source of material properties. Treat them with caution.

INDEX